Suffer the Little Children

I0516756

This book is dedicated to the family of this earth.

Suffer the Little Children

Suffer the Little Children

INDEX

		Page
Introduction	....	5
Soil	....	6
Seeds	....	11
Water	....	13
Clothing	....	32
Shelter	....	35
Electrical Power	....	45
Methods of Raising Water	....	82
Summary	....	101

Suffer the Little Children

Suffer the Little Children

Introduction

With the world population growing in leaps and bounds it may be asked by some why should we worry about the children coming to our congested planet, some even consider that birth control should be imposed upon the population to curb its growth, but when children enter our world are we obligated to help them survive? This book is intended to give a resounding "YES" to this question regardless of where or when they are born.

The title "Suffer the little children" comes from a scripture that is given by the Saviour, Jesus Christ, "Suffer the little children to come unto me, and forbid them not, for of such is the kingdom of God." (Mark 10:14).

Every mother who has given birth to a child can attest that for all the pain, the miracle of birth and the joy of hearing the first cry of their boy or girl, makes the experience worth while. Unfortunately for many in some parts of the world this experience is often followed by the realization that due to the mother's circumstances, their offspring may not live for very long due to disease or starvation.

This book has been written to bring hope to these mothers and to this planet, that there is a way, if adopted, to overcome many of the obstacles that stand in the way of giving every child, water, food, clothing and shelter. This book will also show that there is a source of power that is both non-polluting and available to every part of the world that can give tomorrow's children an atmosphere that is both breathable and sustainable for many centuries to come.

(Diagrams associated with each chapter are shown at the end of that chapter.)

Suffer the Little Children

Soil

One problem that occurs throughout the world is that whereas an area may have been a fertile productive acreage, due to changing weather patterns or natural disasters, these areas have become barren and in some cases desert conditions exist in them.

The question may be asked, can these areas be saved in a natural manner with no artificial intervention?

In North America and in many parts of the world there is a member of the earthworm family that is an ideal partner to man for solving the soil problems of the world. This worm, however, is alien to mankind in its anatomy, in many ways. For instance, this worm has five hearts, no eyes, ears, teeth and moves its body by using bristle power. Its name in Latin is Eisenia Fetilda, (sometimes Foetida), implying foul smelling, as this worm will exude liquid that has a pungent smell, when it is treated roughly. This worm is more commonly known by the names "red wrigglers", "composting or manure worms", "tiger worms" or "trout worms". They do not like human oil from our skin as this affects their own, they do not like our finger nails, hair or skin and so when handling red-wrigglers you should always wear gloves.

What sets these worms apart from other earth worms is that they have the natural ability to eat what we call compost; left-over vegetables, certain fruits, coffee grounds, cardboard, leaves and certain animal manure. As mentioned above these worms do not have teeth and so they feed on the microbes and decaying food associated with such microbes. They then "poop" the most nutritious soil that somehow contains eight times more microbes than they ingested.

Not only is this "worm poop", known as castings, very fertile but when diluted it makes what is called "compost tea", which may be used as an excellent fertilizer for lots of plants.

Suffer the Little Children

There are many theories on the lifespan of the "red wriggler" but some say it is one month and others as much as ten years. However, either estimation is hardly relative when we realize that the worm population can double every 11 weeks; if fed properly and that they have enough space to expand that population. Each worm can eat double their weight every three days and thus the amount of castings they produce is staggering. Two pounds of worms, about 2000, can create about seven pounds of castings within a month. (Since they only grow to four or five inches that is an impressive amount of castings.)

The red wriggler is in the category of a hermaphrodite, where it has both male and female sex organs, but it does have to have a partner to have sex. When the worm is sexually active you will observe a ring around their body near to their head, known as the clitellum. They then find a partner that they go head to tail with and tightly bond to each other using bristles on their underside, (called setae), secreting a seminal fluid. The worms then secrete a mucus ring around them, which begins to harden before the worms slip out of it. The worms pass the mucus ring over their head and as they do the ring picks up the sperm to reproduce. Each cocoon can contain up to twenty eggs although only two to six is average but even so with an eleven to twelve-week maturity window the population growth is quite rapid. (Usually only about three worms hatch from one cocoon.) The cocoons are easily identified as they look like small, pale yellow lemons but as the worms grow within, they change to a brownish colour. The worms enjoy temperatures between 60 – 80 degrees Fahrenheit, and the cocoons usually hatch if that temperature is maintained. When the temperatures drop or rise above this parameter then the cocoons may delay their hatching period. It appears that this is one method that the red wriggler survives harsh weather and thus a worm bin is seen to have many babies hatch in the spring to compensate for the loss of more adult worms. The hatchlings do not begin as a red colour but are pale, almost white, but this quickly changes to the distinctive darker red colour. Of interest is that when a worm is born it has 120 segments (rings) that are moved by tiny circular muscles beneath their skin. These 120 segments then expand as the worm grows so regardless of its size it always has the same

Suffer the Little Children

number of segments. They also secrete a slippery fluid that with the underside bristles (setae) help them move.

Unlike other earth worms they do not like to burrow much below six to twelve inches and thus many worm farmers take advantage of this fact by feeding them from the top and then as the colony moves up leaving the rich castings below them, it makes harvesting the crop of castings that much easier. The red wrigglers do not survive long in regular soil, but then other earth worms do not survive long in the red wriggler castings either.

Summarizing then we have a creature that is not only beneficial to mankind, by providing nutritious soil but they are almost independent in their operation to do so. When leaves are used as bedding for them it acts both as a place for them to live and also becomes a food source as the leaves shrivel and become mushy when damp. As has been mentioned the red wriggler enjoys food scraps but there are some things that they do not like, including anything from the onion family, meat, dairy products, citrus fruit, cooking oils or grease and definitely, they do not like human or pet waste. Some of the largest worm farms feed their worms on a diet of cardboard slurry which contains corn starch. (Interestingly, if you wish to stimulate sexual activity in a worm colony add small amounts of corn meal.) It has been found that red wrigglers love members of the melon family, especially water melon. The worm bin should be damp but not saturated as your worm colony will happily leave such an environment. That is why a worm bin should be large enough to allow plenty of movement for the worms, so that if one area is too wet, too hot or too cold, they have another area to move to. They also like plenty of air circulation but this is usually not a problem if the top of the worm bin is open.

Although there are many forms of worm bins, (many can be found on the internet), it will depend on your demand for soil as to how big of a bin or bins you will require. One hint that I learned, if you decide to make your worm bin from wood, instead of painting, it is better to use a paint brush to brush the inside of the bin with cooking oil. Worms will eat wood but by coating it with the cooking oil, it not only preserves the wood, but as the

Suffer the Little Children

worms do not like cooking oil they will stay away from the sides of the bin.

Another method of storing worms would be to use bags or plastic tote bins. The bag method should allow you to use any bag; from industrial garbage bags to polyethylene woven bags (as in sandbags), to house the worms. At first sight it may appear that an industrial garbage bag would be an odd choice since it has no drainage, but if the contents are not saturated and the opening to the bag is unobstructed the worms seem to love to hug the black plastic when not eating. Plastic totes will work for projects that allow a few worms, (around 100), to occupy each tote, however, such bins need a lot of ventilation, temperature control and measured food supply to be effective.

When liquid passes through the worm bin it emerges as a brown liquid known as "leachate" which can be used as a form of compost tea.

Thus, we come to the production of the fertilizer known as "compost tea" and how to prepare it. Take some worm castings and put them in a sack, before placing it in water. Allow compressed air to pass through the liquid for at least twenty-four hours to aerate the mixture; to which molasses (not containing sulphur) is applied. The resulting "tea" is rich in microbial activity that plants love and allow them to flourish. One of the many advantages to this "compost tea" is that it is 100% chemical free and therefore will not harm the roots of tender plants. However, many farmers dilute the mixture further to extend its use on the crops. I personally like to put the worm castings into a twelve-litre container (usually about one pound) to which I add eight litres of rain water. I then add one teaspoon of black strap molasses (no sulphur) and then run air through the liquid (using an aquarium pump) for twelve hours. I then strain this mixture through layers of cheese cloth sandwiched between two large strainers. The resulting compost tea is very rich in microbes and nutrients, therefore giving a good ph-value to any soil.

When harvesting the worm castings, there are several methods shown in different literature. One of them is making a small pile and casting light on it, (since worms do not like light), thus sending the worms to the bottom, allowing you to carefully take

Suffer the Little Children

away the castings above them. This is fine for small amounts but commercially there are mechanized strainers that are cylindrical. When the motor turns the device, the holes in the strainer are such that castings fall from one side and worms are collected on the other. I prefer to use a milder way of making a strainer or sieve that has holes ¼ inch apart. I manually put the castings into the sieve and by moving the castings around, I am then able to extract the worms without inflicting the stress of a machine upon them and they are quickly put back into a leafy environment. A lot of worm farms also use winnowing, where a plank below a narrow trench, is drawn along mechanically thus taking the lower castings and dropping them onto the floor below the trench to be collected manually later. (The internet shows such operations in more detail.)

Some worm farms store the castings for up to a year; ensuring that it is at its peak nutritional value. Other farmers bag the castings immediately and send them to retail outlets or garden centres. Since the intent of this chapter is to provide soil for the world, the bagging method would appear to be the best.

Although we have spoken a lot about the red wriggler, there is another earth worm in hotter climates that also acts in the same way, called the blue worm (Perionyx excavatus). This blue worm would therefore be ideal for countries such as Africa, India and Central America as well as other such warm climates.

Having now found the perfect soil and fertilizer for growing plants we will now consider what we would want to grow and how to provide seeds for that growth.

Suffer the Little Children

Seeds

There are many seed companies that will gladly sell to you seeds online or in stores but as we are talking about feeding the world such outlets are not always available to everyone. One rule of thumb is that you should check out your local area to find out what grows there already. Although in North America the main crops appear to be corn, soya and wheat there are many other fruits and vegetables grown for local use. In Asia there is a tendency to grow rice but again there are many varieties of fruits and vegetables available to their markets. Some countries have fruits that are peculiar to their locale such as the kiwi from New Zealand. Some equatorial countries are famous for their pineapples, bananas, coconuts and cocoa beans. Although these products are shipped all over the world, they may not be available to areas of concern where children are starving today.

The last chapter has shown how to provide all the soil the world needs but what if we wish to grow our own local fruits and vegetables from scratch. Nature has provided a method of supplying such a need with seeds. It never ceases to amaze me how an apple seed knows to grow to be an apple tree instead of a pear or other tree, yet within the tiny apple seed is all the chemistry necessary to grow apple orchards. To provide yourselves with seeds for next year's planting is as simple as taking the seeds of a fruit or vegetable, carefully washing off any material that could rot the seed, dry the seeds for a few days, package them in freezer bags and freeze them until the following spring. Seeds are so important that in Norway there is a giant vault that is heavily guarded where they store seeds from across the world: (Svalbard Global Seed Vault).

Some plants have been altered so that they do not contain seeds, (such as bananas, grapes and some melons), and so these, although popular could cause a shortage of such crops in days to come.

Grains such as wheat, oats and rice are very important to one's diet and so knowing how to grow them is essential. Wheat can

Suffer the Little Children

be grown from the kernels and are so resilient that some wheat found in mummies' tombs, have been planted successfully, 4,000 years later than the time they were placed in the tomb. Wheat, however, does not start as a grain but in its early stages it is a vegetable known as wheat grass. In this stage the wheat has amazing nutrient value, containing many vitamins, enzymes, minerals and other chemicals essential to good health. Although the wheat grass needs to be ground down to extract the juice, (since we do not have the stomachs of cows), this juice, when fresh, is of great value to our body: building our immune system and many other advantages, (which can be explored on the internet.)

Although oats are associated with feeding horses, many nations use it as a breakfast cereal, "oatmeal or porridge," as it is high in fibre. Rice has been the staple diet for many, probably about a third of the world's population, for thousands of years, and although it does take a lot of water to grow, it is not hard to grow in the right conditions.

As has been mentioned before fruits and vegetables that are common to any certain area should be the first choice for growing, but one European staple; turnip, is one vegetable that is grown widely throughout the world, in various climatic conditions, and shows a resilience to many diseases. Another favorite is the humble potato that is grown in many parts of the world and has become a staple for many countries. Interestingly, the potato grows well in compost.

The climate appears to be making desert areas appear where they have not ben seen before and some areas have even lost any soil that had existed before. This puts a heavy strain on communities who had relied on such areas for food but one method that may help in this situation is what I call trench farming. However, before we go into more detail about this we will consider the other major obstacle that some are feeling and that is drought or no accessibility to water. The next chapter will cover this problem and will hopefully provide the solutions to them.

(Of interest, Israel and China have been making great strides in growing crops in both salt water and in desert areas.)

Suffer the Little Children

Water

Although our planet is made up of over 70% water most of this is sea water and thus not very useful, as such, in agriculture due to the large salt content. In this chapter we will discuss desalination, how to transfer water on level surfaces, how to sterilize polluted water and of major importance, how water is to be used in trench farming.

The usual way that water from the sea falls as rain is that the sun evaporates the sea water, allowing it to rise as vapour into the atmosphere, where it then condenses into water droplets that eventually form clouds. As these clouds become saturated they gain enough weight to fall as rain, snow, sleet or ice depending on the ground temperature. However, although some rain falls onto dry land there is a lot that falls back into the sea. In fertile areas like those around the equator this rainfall can be extensive and cause a lot of vegetation to grow but in desert areas there is the opposite: a lack of moisture, thus whether it is a desert like the Sahara or one like the Arctic, the problem arises that little moisture means little or no growth of plant life.

Israel has many areas of desert and has solved much of their water shortage with desalination, using filters and artificial evaporation methods to provide clean water for drinking and irrigation. In the desert regions they then provide thin pipes that are on the ground and have small holes at certain points, so that the water drips onto the ground. Plants are placed at these drip points and thus the roots are nourished enough without extensive evaporation from the sun's heat. Through these methods, Israel has now an abundance of water and many crops to feed their people.

My idea of trench farming would be ideal for similar situations but particularly where the surface soil is poor or non-existent. Fig. 1 shows a network of trenches, each twelve feet deep and twelve feet wide. The trenches are dug in two stages; the first trench will be twenty-four feet wide, six feet deep and one hundred feet long. To support the walls, the "Earthship" method of filling used tires with earth, will be used so that one row of tires, (see Fig.

Suffer the Little Children

1a), will be built on each side of the trench. Stage two will be the digging of the second trench, within the first, but this one will be sixteen feet wide, another six feet deep, (making twelve feet from the surface), and the one hundred feet length. Again, the "earthship" method of used tires will be used but they will form a two feet wall in the bottom trench and a four feet wall on the upper trench, thus leaving a trench that will be as illustrated in Fig. 2.

At every twenty-five-foot interval, a column arrangement, as shown in Fig. 3, will be built. It consists of a pillar, with mirrors, to reflect the sunlight from above the trench into the trench. This provides both light and sunshine to make the plants grow. On either side of the column is a framework that will support a corrugated plastic roof, that will be covered in dirt, so that the only light entering the trench will come from the columns.

The walls of the trench will be lined with cement, (or a local equivalent), and the floor will look like the one shown in Fig. 4. This shows that the floor will slope inwards to the centre where there will be a channel for water to flow. There will be a pathway for workers to walk along the channel and between the plant beds. The plant beds will also slope towards the channel and be constructed so that the soil will have drainage.

Fig. 5 shows the tanks where water is stored for irrigating the plants. The method by which they will be filled will be discussed later. There will be a tap (faucet) on each tank

to allow the worker to either connect a hose or use a bucket for the watering process.

Fig. 6 shows that at the end of each trench there will be a building so that workers can enter and exit by them, air can circulate through the trenches and the building will be raised to keep out unwanted guests, such as snakes, etc. entering the trenches uninvited. A ladder will be used for the workers to enter the trench and a pulley system may be used to lift heavier harvests.

The trench farm will then not be subject to intense heat or unnecessary insects or other intruders. The advantages will be

Suffer the Little Children

that many crops can be rotated as there will be an abundance of water and sunshine, the food being grown locally can feed the people around the trench farm and excess can be sold or bartered with other neighbourhoods. The soil, coming from the vermi-composting areas previously described, (red or blue worms), along with the compost tea will provide all the nutrients any plants will need. The water will add to that nutrition, through the plants' roots, and the sunshine will give the necessary light, thus the plant will have all the potassium, nitrogen and phosphorous associated with usual fertilizers.

Since the trench farm can be expanded, as shown in Fig.7, regardless of the size of the demand, food can be grown to meet it.

Coming back to the title of this chapter: water, we will now take a typical situation in a possible African environment. The picture is shown of a person carrying water on their head for miles at a time, collecting the meagre supply from often polluted water in an almost dried up well or other hole in the ground. There are many prevalent rivers in Africa, but they are not always convenient to every village. Another problem is that the terrain may be uneven and since water can not run uphill without pumps, to transport water long distances may be impractical.

The solution may lie in the use of siphons. In other books I have described how Pythagoras was given a challenge to create a cup that would not let a person drink too much wine. His solution was to create a column in the middle of the cup which hid a siphon such that when a person took the right amount of wine the siphon was inactive but as soon as they exceeded that amount the siphon would kick in and the person would find wine all over their lap. (see Fig. 8). This cup, known as the Pythagoras cup or the greedy cup is still sold as a novelty in Greece.

Although the siphon is usually shown with a short bend between the short and the long end, it has been found that the bend may be extended for sizable distances, and so long as the siphon has a short and long end the action will still work. This being the case the arrangement shown in Fig. 9 will work, where the short end is in a container and the connecting tube is kept level until it reaches the other container where the long end of the siphon is.

Suffer the Little Children

As with the Pythagoras cup, so long as the water level in the first container is above the tube the siphon will work until the water level of both containers is equalized. If we then add a third container with the short end of a siphon being in the second container and the long end being in the third container, again so long as the water remains above the tube, the water will equalize in all three containers, (Fig. 10). The important but maybe not obvious fact is that there needs to be air above each container; as the air needs to be replaced by the water.

Taking this example further, if the tube is one mile long, the method will still work so long as water in container one is above the tube. Fig. 11 shows container two having a hand pump that also allows air to enter the container, thus if water is pumped from container two, so long as container one is full then as water is pumped out, container two will refill itself. This concept can then be expanded to several containers, one mile apart, each having a hand pump and if we assume that container one is full, each container will continue to refill itself. It will also not matter how deep the containers go so that in our trench farm system the tanks can be twelve feet deep and they will still refill; if we ensure that container one is full, and the tubes are kept level, (see Fig.12). The village may also have a central container that will act as a well and save persons having to trek miles to carry a meagre amount of water.

We now come to how to maintain a constant supply of water in container one. This is illustrated in Fig. 13. where a building is constructed next to a river or lake. Water is diverted through filters to an artificial lake and this water is then fed into pipes that are heated by a fire, thus creating steam. This steam then condenses within a mineral rock silo that will then feed container one with distilled water laced with mineral salts, that will be drinking water. The fire will be provided by either local wood or using the dried dung of animals or humans. Obviously, this facility will have to be large and thus a whole village may be needed to man it, being paid by the recipients of the water in food from the trench farms. Fig. 14 shows a schematic of the siphon system.

Suffer the Little Children

Although it would appear at first that this project is ambitious, it is not impossible. The initial trenches to carry the tubing, the construction of containers with pumps, the building of a water purification plant will also provide labour for the villages involved. It is also possible that the heating of the pipes in the water purification can be provided by electricity from either wind or solar power, but it will depend on the location of the river or lake to see if that would be feasible.

Thus, if this method is adopted in all areas of the world where rivers and lakes are available as a source of water, there should not be a shortage of water anywhere. The challenge, however, may be to keep the tubing level, since this may mean traversing valleys, gorges, mountains, hills and other obstacles but again it is a feasible solution. (Note that if chamber one has a three-foot head it will allow for a three-foot clearance for the tube to vary from its level position.)

We have mentioned above that desalination plants are a way of providing clean drinking water to the world. Environmentally there is a concern that the salt left over from desalination may upset the balance of the sea's ecosystem, but it is possible that if the salt was taken farther into the sea and scattered over a distance, this will not be a problem. The word desalination implies the removal of salt and as mentioned above Israel has made great strides in modern desalination methods, using filters as well as heat to distill the water. One drawback at present is that the process requires a lot of electricity, however, as will be explained at the end of this book that will also not be a problem if the world converts electrical generation from fossil fuels to the use of compressed air and water.

The earth has many aquifers, (an aquifer is an underground layer of water-bearing permeable rock, rock fractures or materials such as gravel, sand or silt, from which groundwater can be extracted); some being large and others small localized ones. Unfortunately, many are in danger of pollution from various chemicals used throughout the world, and some are in peril of drying up due to over-usage or lack of moisture to replace the supply taken from them. Mankind must treat these aquifers with a lot more respect than they presently do, or there may be dire

Suffer the Little Children

consequences for the future of our planet if they do not. It is the same for rivers and lakes which are being bombarded with chemicals and acid rain that also can not be tolerated. The answer is a global clean up that may cost a lot of money, but unless mankind takes responsibility for its wanton destruction of earth's ecosystem, our children and grandchildren will not be pleased with the world's legacy they have been given.

Water is what sets our planet apart from all the others in our solar system, therefore we need to be aware that it is precious and must be used with great care.

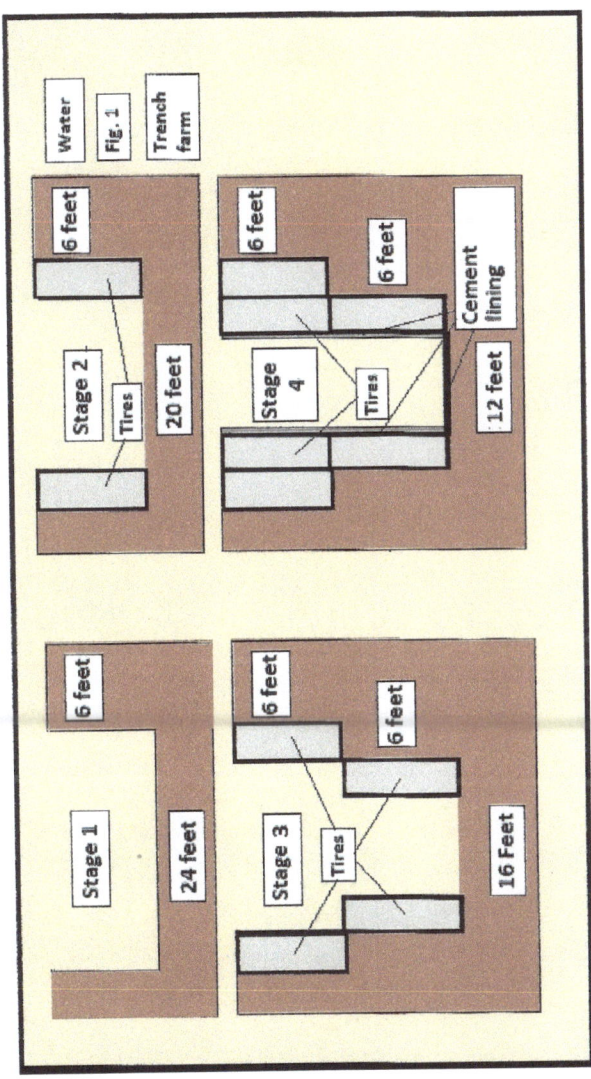

Suffer the Little Children

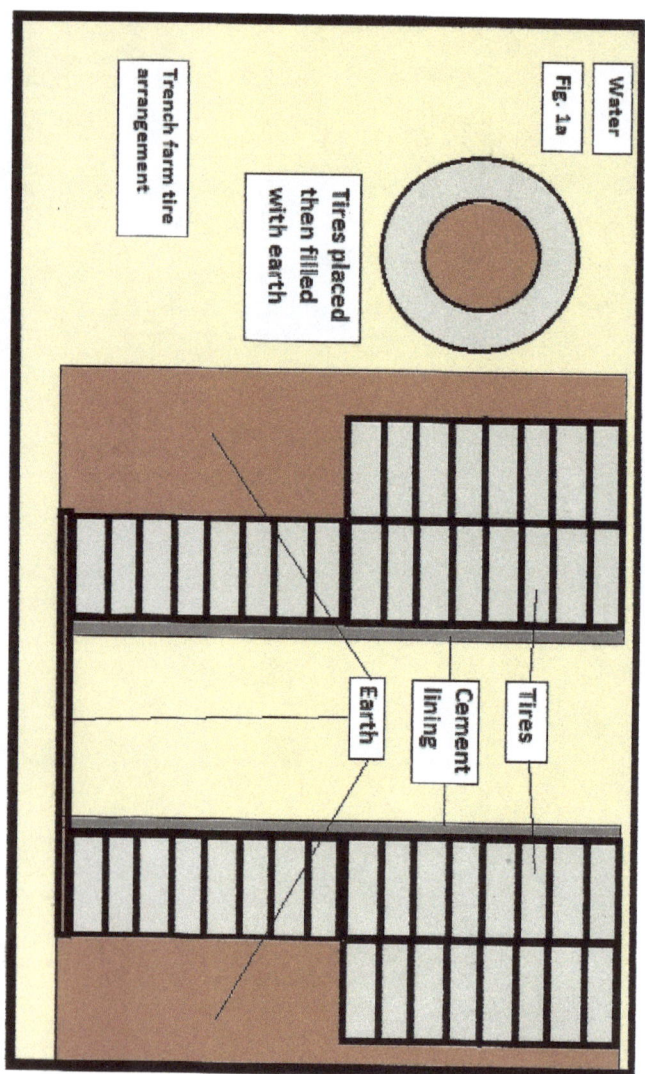

Suffer the Little Children

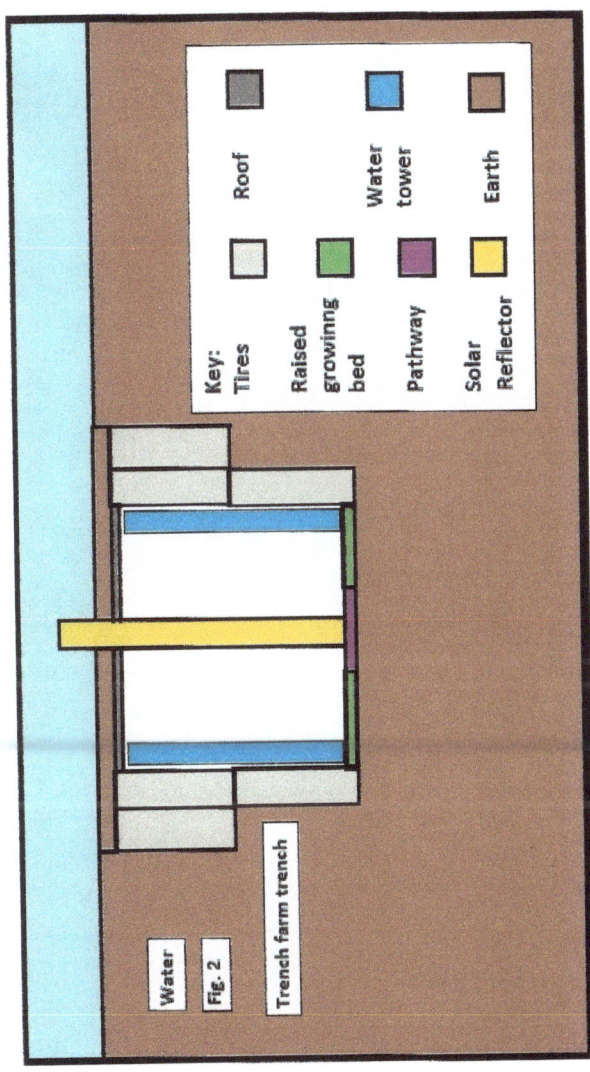

Fig. 2 — Trench farm trench

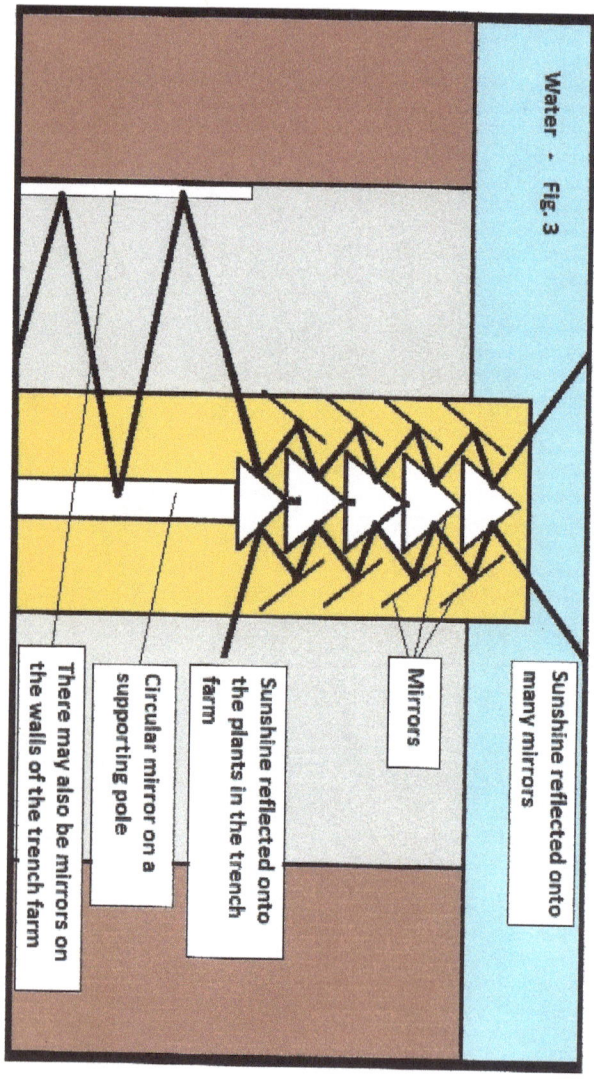

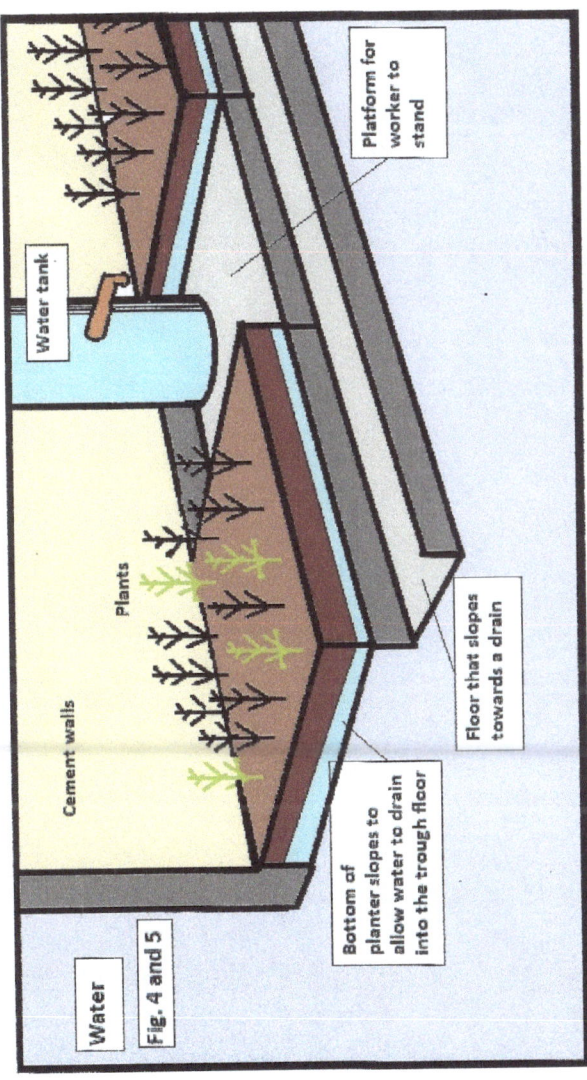

Suffer the Little Children

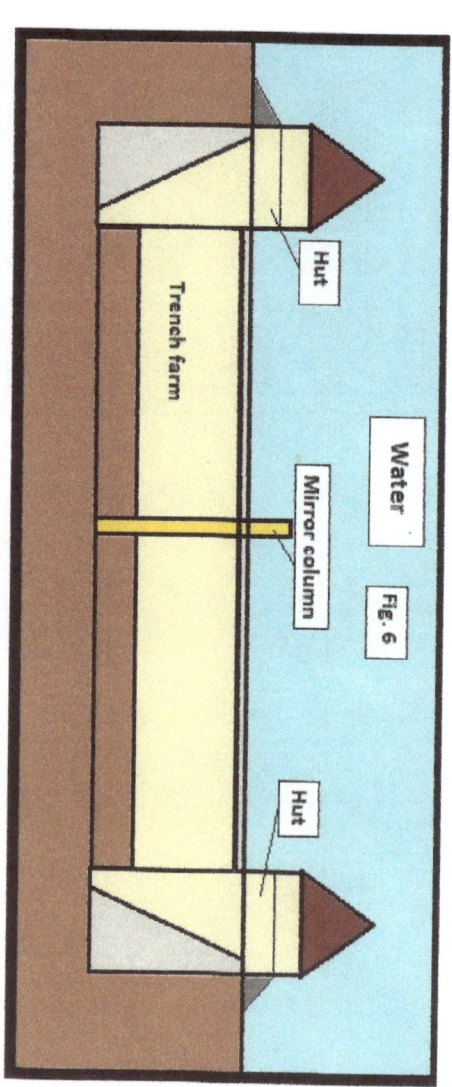

Fig. 6

Suffer the Little Children

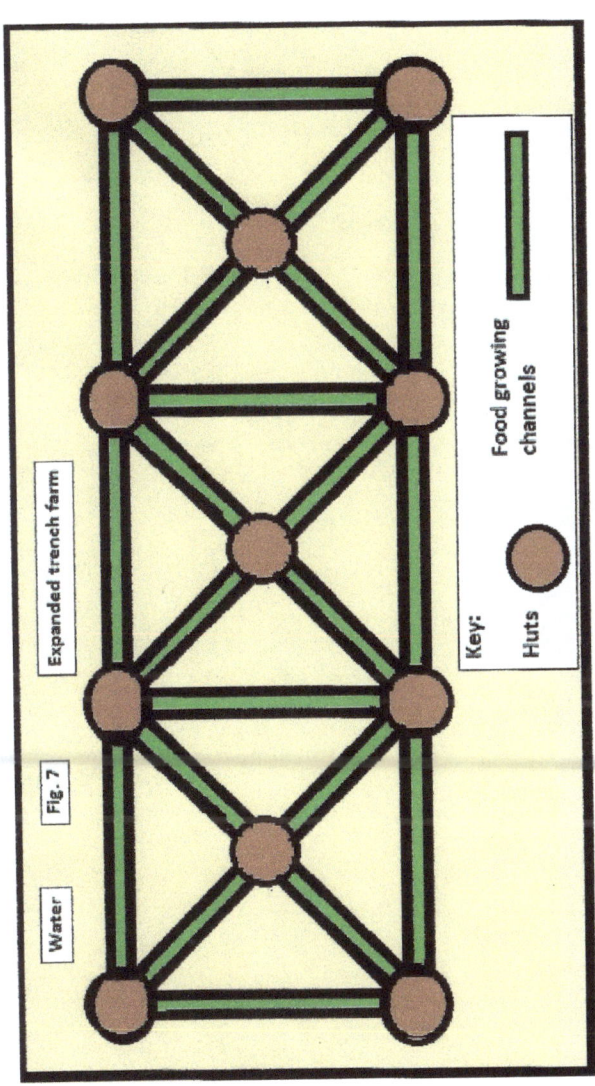

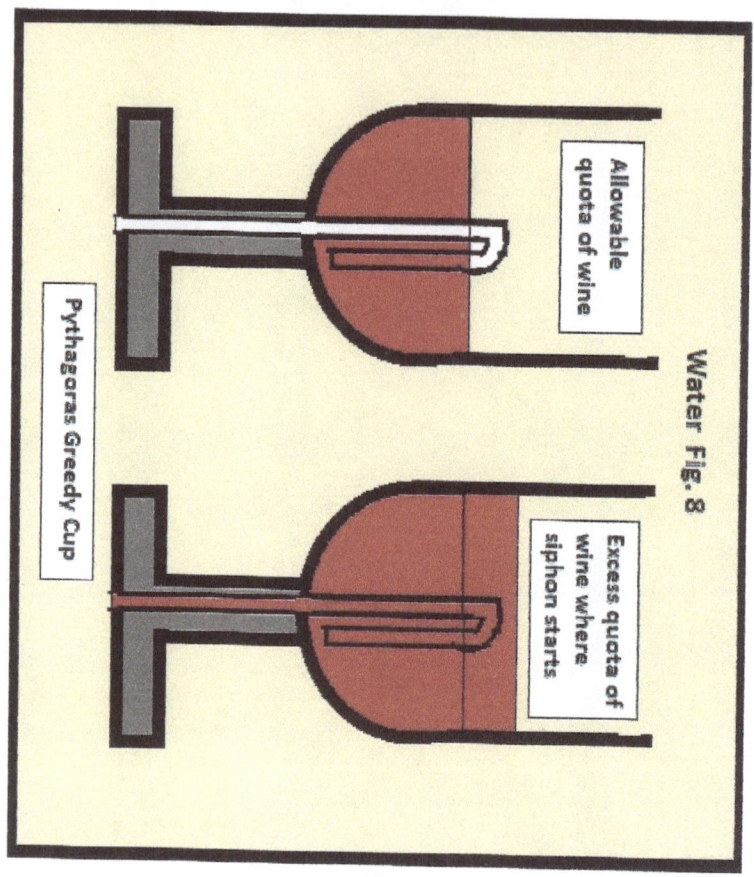

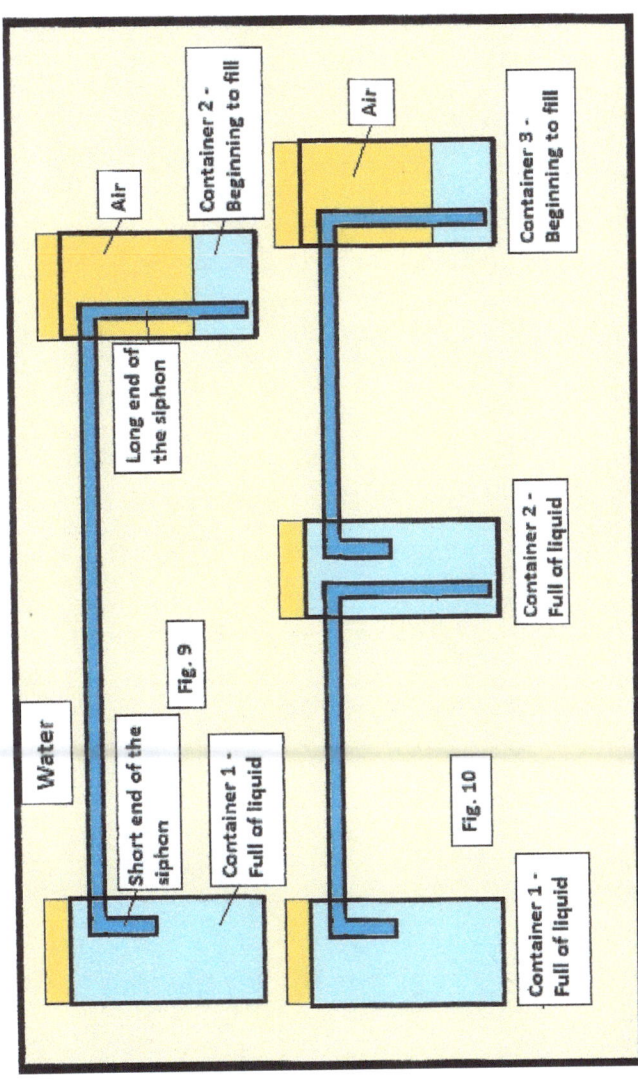

Suffer the Little Children

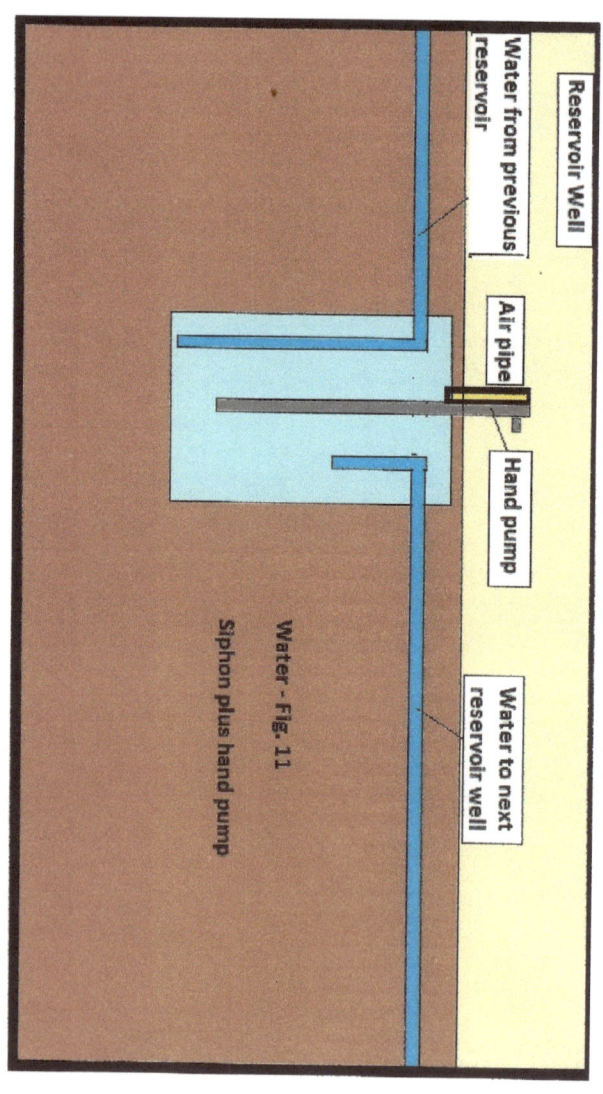

Water - Fig. 11
Siphon plus hand pump

Suffer the Little Children

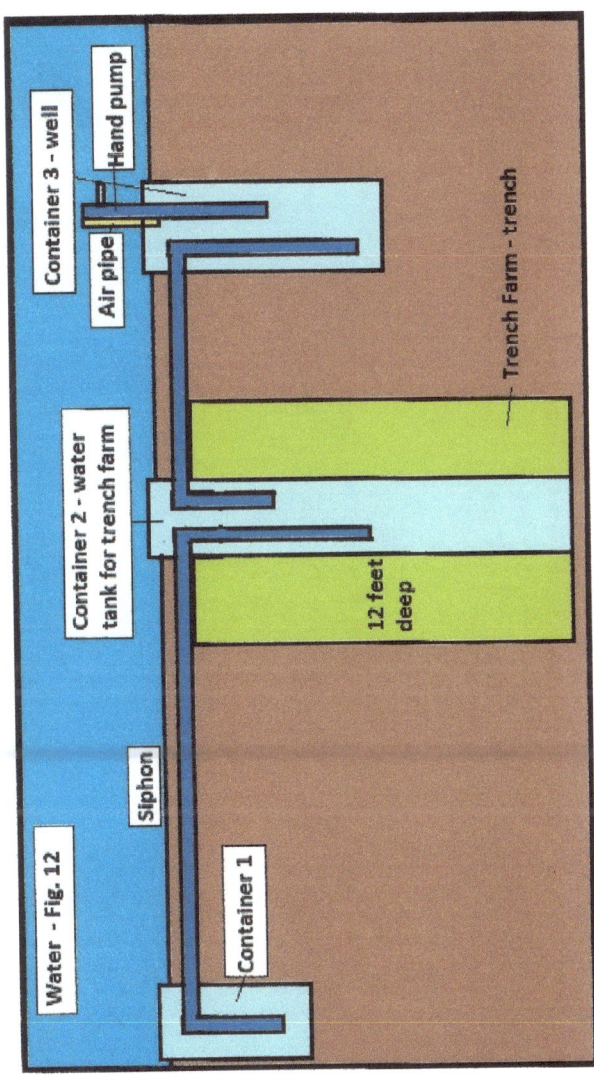

Suffer the Little Children

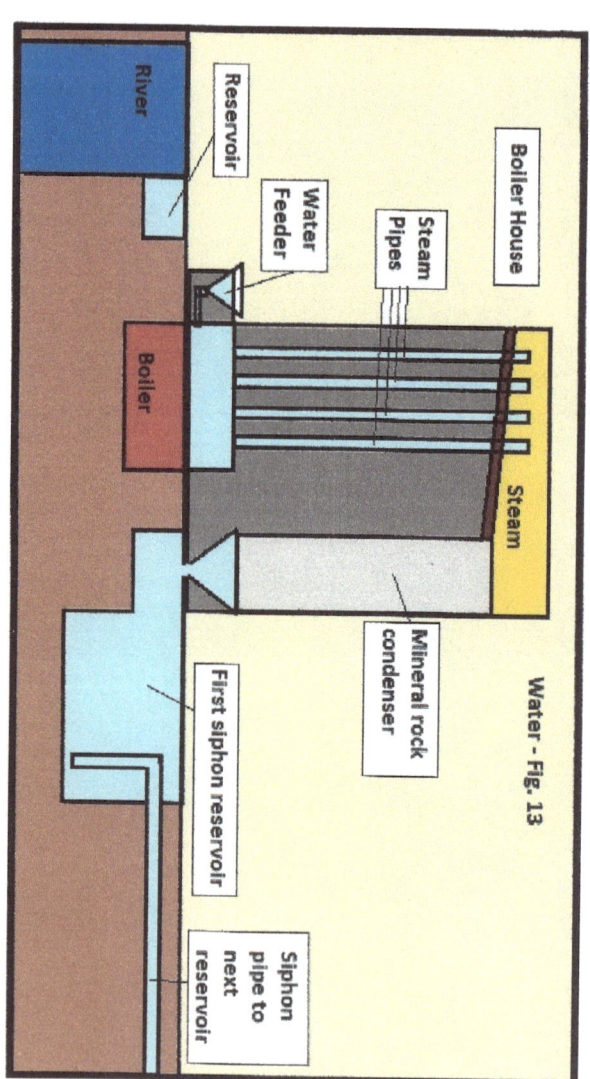

Water - Fig. 13

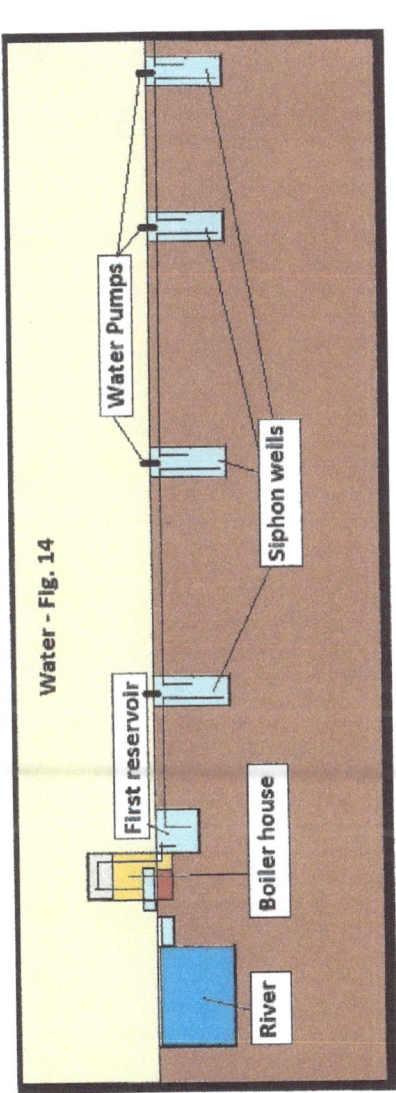

Water - Fig. 14

Suffer the Little Children

Suffer the Little Children

Clothing

Since the dawn of time mankind has used many materials for making clothing. From animal skins to woven threads, man has continually expanded their knowledge of fabrics and today we are blessed by that knowledge with every conceivable material being made into clothing. Whereas, a century ago clothes may not have lasted long, today there are materials that are almost indestructible.

Two skills that have developed over time has been knitting and weaving. The industrial revolution of the nineteenth century in Britain brought about the mechanization of weaving on a large scale. Cotton became the main commodity of the day and factories sprung up all over the United Kingdom, particularly in the North of England. In Scotland, (Dundee), jute became the major weaver's preference, but this cloth was used more for sailing ships and the covered wagons of the American pioneers. Silk weaving was also present in London, England but that was also taken over by mechanical looms. There are still hand looms and they are used mainly for recreation, but the art of weaving is one skill that will never go out of style.

Knitting, whether plain or more complicated stitches, has also been duplicated by machinery, but there are many who still use the old-fashioned knitting needles to create clothing in all kinds of designs. Some prefer the man-made fibres, such as polyester, nylon, acrylic or other such "wool" substitutes, while others prefer the natural wool of sheep to create their delicate but long-lasting wares.

Both skills are essential to the well being of any society, combined with sewing, to develop any clothing that is necessary to that society. Therefore, we would expect that these topics, being important, would be taught to children as part of their education, but sadly this is usually not the case. Instead there are thrift stores all over the civilized world that except donations in one door; to be displayed on racks and leave by another door, having made some charity richer and creating jobs for workers.

Suffer the Little Children

In this way the flood of new designs, fashions and fads that come into vogue and seem to leave as quickly as they arrived, are able to be recycled.

For the starving children of the world, the lack of clothing seems to follow this part of society and the parents of such children can hardly afford even the clothes from thrift stores, if they were accessible. There are world charity organizations that collect used clothing and send it to the under-privileged, but their resources are usually limited and with an increasing world population the demand far exceeds the supply. One solution would be to teach the skills mentioned above and then instead of providing ready made clothing, raw materials, which may weigh less and are less bulky to ship, may be sent to these countries allowing them to create their own clothing and then with such skills, be able to barter or sell to others to start to make a living.

The demand does not only stop at clothes, shoes are another commodity that are sadly lacking in many areas of the world. Although there are many places where shoes are not necessary, in those areas where they are, again the teaching of basic cobbler or shoe maker skills, would help.

Thus, the adage of Confucius, "Give a man a fish and he will live for one day, but teach a man to fish and he will live for a lifetime", becomes pertinent to the situations described above.

The term clothing can be more broadly applied to other essentials, such as blankets, ponchos, quilts and other bedding requirements. Again, the skills of weaving and knitting can come into play to make such items.

If possible, animals could be reared to provide wool or yarn of some kind, such as sheep or lama, and could become part of the farming environment that would be created. The skills of shearing, carding and other steps towards the production of yarn or thread would be taught by outsiders to allow the under-privileged to acquire more knowledge for their own subsistence and to possibly provide a means of employment and therefore money.

Suffer the Little Children

The principle of self reliance is essential to the well-being of all people upon this earth and through concerted efforts this may allow the poor of this world to lift themselves out of their poverty situation and into a place, where families can be reared in reasonable circumstances.

Suffer the Little Children

Shelter

From the primitive cave to the stately mansion, mankind has found many ways to put a roof over their heads. Throughout time man has discovered different building materials that allow the building of more ambitious structures. Dubai, capital city of the United Arab Emirates, is a perfect example of man's ingenuity to build a metropolis in a desert region, which is still expanding. We can thank many nations for our building materials, from the common house brick to the cement mortar. (Many such items came to us from as far back as the Roman Empire.) Iron and steel have allowed humans to build sky scrapers many storeys high. Tempered glass and the introduction of plastics has allowed windows to come in all shapes and sizes. Mankind has built high buildings and in colder climates have succeeded in burying themselves underground; creating shopping malls connected to each other so that they do not have to go out into winter's storms, unless they want to. Underground railways have been built to transport people from home to work, again defying the cold elements that Mother Nature has given us.

Although there is concern that soon the world's population may reach 9 – 10 billion people, there are many areas of this earth that are uninhabited. Canada, the second largest country in the world, has vast expanses of land that are not inhabited, mainly because of climatic conditions, but with ingenuity these problems can be overcome. It is hoped that man may one day inhabit Mars, so if we will be able to tame such a planet it should be a breeze to tame the arctic terrain of Canada. The Sahara Desert spans much of Northern Africa and is the third largest desert, next to the Arctic and the Antarctic. If the Sahara could be tamed, that would give us another vast expanse to work with. However, the largest area of the world, yet untapped, is the seas and oceans that surround our land masses. There are many science fiction writers who design underwater cities based on the legends of "Atlantis" but they usually do not take into account the fact that the deeper you go in our oceans, the greater the pressure is exerted on any structure.

Suffer the Little Children

What if there was a way to overcome such pressures? There is: as I discovered using an ordinary bleach bottle. If a container has a bleach bottle in it, with a hole in the handle of the bottle, as in Fig. 1 and water is poured into the container; so long as the hole is under the handle, the water will not enter the bottle. The reason for this is that the air pressure in the bottle is sufficient to hold the water out. However, although this air pressure is minimal because the handle is under a small amount of water, that same pressure is consistent throughout the bottle and thus even though the pressure of the water is greater outside the bottom of the bottle it is still the same in the bottle so long as the sides of the bottle can withstand the outside pressure. If we now apply this principle to an actual structure, (see Fig. 2), you will see that a tower has been built in a body of water. The base of the tower is much thicker than the top as the water pressure will be quite substantial at the base. The top of the tower is capped by a dome-like structure that is shown to have two air vents. Compressed air is then blown into these vents at the pressure of the water at that point of the building, (ie: if the air vents are ten feet under water the compressed air will be such that it will compensate for that ten feet of water pressure.) Note that the air is not pumped directly into the vents but into the water below it. This will mean that should the compressed air be shut off for any reason the water will not flood the tower through the air compressor pipe. As the air fills the tower a point will be reached where the stale air may be sucked out by fans and the other two air vents will be used to channel that air into the water and back to the surface, thus at no time is there any chance of carbon dioxide build up once people start to inhabit the tower.

Fig. 3 illustrates a concept of a typical tower with several floors, some to be used for dwellings, others for stores and others for schools, hospital, church, gyms and other recreation areas. Thus, the tower becomes a self-sustaining building for several families. Fig. 4 depicts the concept of a group of these towers connected by a surrounding wall and five extra towers within the compound, three to provide electrical power and two to provide agriculture. The electrical power will be discussed at the end of this book, but the agricultural towers will be described below.

Suffer the Little Children

Besides providing food for the colony, the agricultural towers will also house some livestock, such as dairy cows, goats, sheep, chickens and pigs. Their manure will be processed by vermi-composting, (as covered previously), to provide the soil for growing fruits, vegetables and grains. The sewerage from the human towers will be shipped to the agricultural towers, dried and used as a possible heat source for the human towers. Obviously, skill will have to be developed to make sure that odours and methane gas do not hamper the operation of the agricultural towers. Fig. 5 shows the envisaged agricultural tower. Note that the use of submarines will be used extensively in the colony to ferry supplies, food, provisions and people from one area to another.

As mentioned the colony will be surrounded by a wall, which also will have a thicker base to withstand the water pressure. When the colony is to be built, the wall will be constructed under water so that the area of the colony will be created. The human towers will take the wall as one side of their structure as shown in Fig. 6. The water will then be drained from the enclosed space so that the construction of the colony may be accomplished in a dry environment, meaning that the wall will be high enough to keep the water out. However, there will be a series of holes drilled partly into the wall about twenty feet from its top. This will mean that the walls will still be able to withstand the water pressure but when the structure has been completely built these holes will be drilled so that water will enter and fill the compound. These holes will not be very large and thus if there are fish in the surrounding water, their young will enter the compound via the holes and as they grow, will become another food source for the colony. This also means that predator fish, such as shark, etc. will only be able to enter when young and thus scuba divers in the colony will be able to extract them and return them to the outside before they grow to any significant size.

It is estimated that a colony will house about 15,000 people, (men, women and children), and the agricultural and electrical towers will sustain this number, supplemented by the fish supply in the colony water. It can then be seen that if several colonies were built in lakes, seas or oceans there is limitless possibilities to providing shelter for the world's ever-increasing population.

Suffer the Little Children

If the colony is in a sea, then it will be expected that storms will strike them from time to time, but it will not affect them as the towers will not flood so long as the entrances are under water. The only effect the storm may have is to slightly increase air pressure, but this should not be a problem for the inhabitants of the different towers.

The inhabitants will use submarines to move between the human towers, but the use of scuba diving equipment should be the major gear for most colony dwellers. This means that the colony will create a new world for these people as they will probably spend more time in the water than on dry land.

Fresh water for the colonists in the sea colony will come from desalination and be shipped into the human towers by submarine.

This concept can be built on land and then the area could be flooded to create the same colony effect.

Thus, it would seem to me that regardless of the size of the population of our earth, there are ways and means to give everyone shelter.

Suffer the Little Children

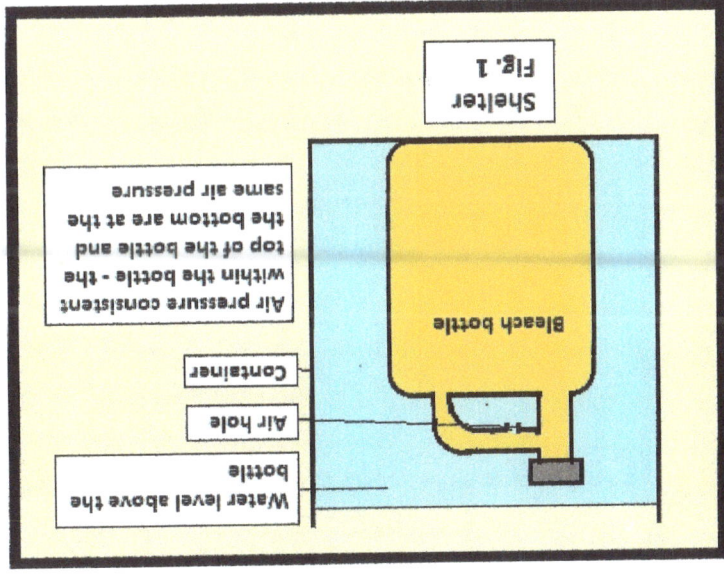

Fig. 1

Shelter

Bleach bottle

Air pressure consistent within the bottle - the top of the bottle and the bottom are at the same air pressure

Container

Air hole

Water level above the bottle

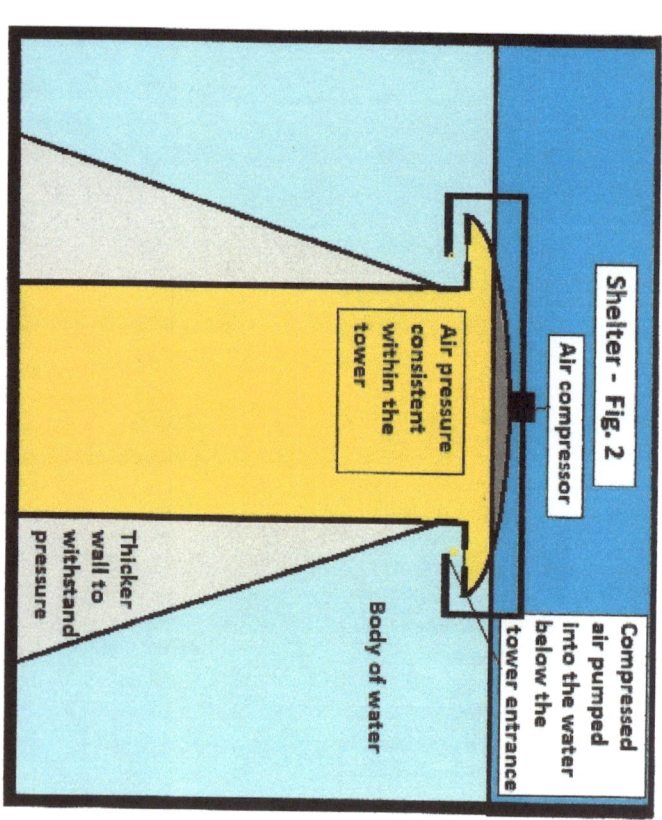

Suffer the Little Children

40

Suffer the Little Children

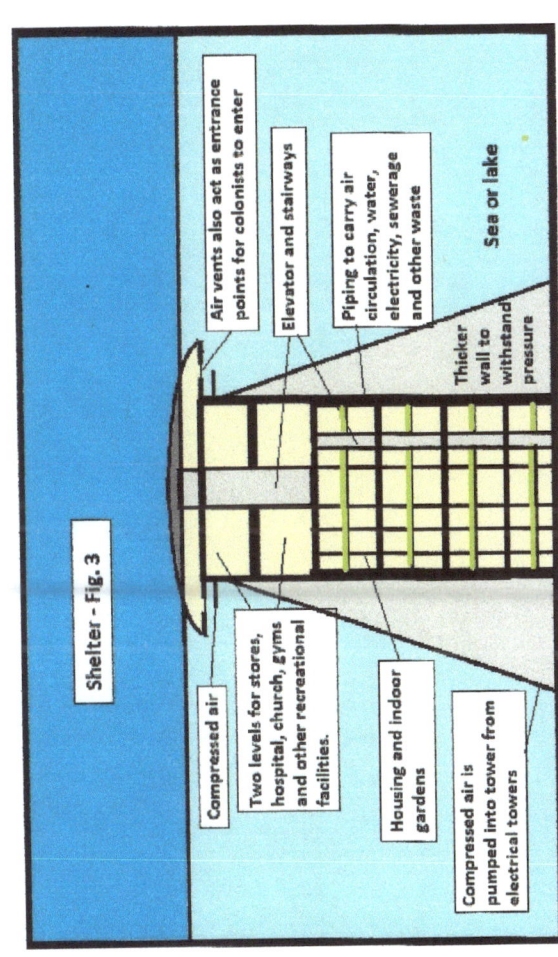

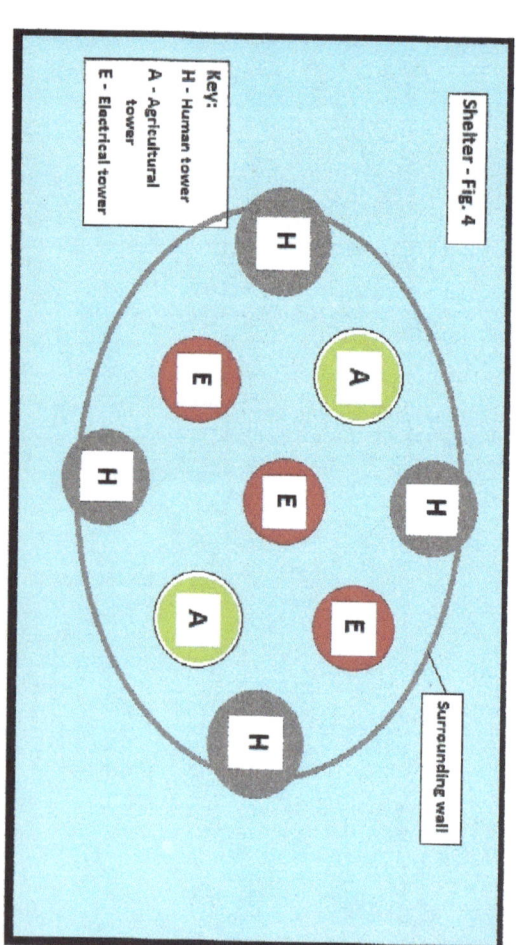

Shelter - Fig. 4

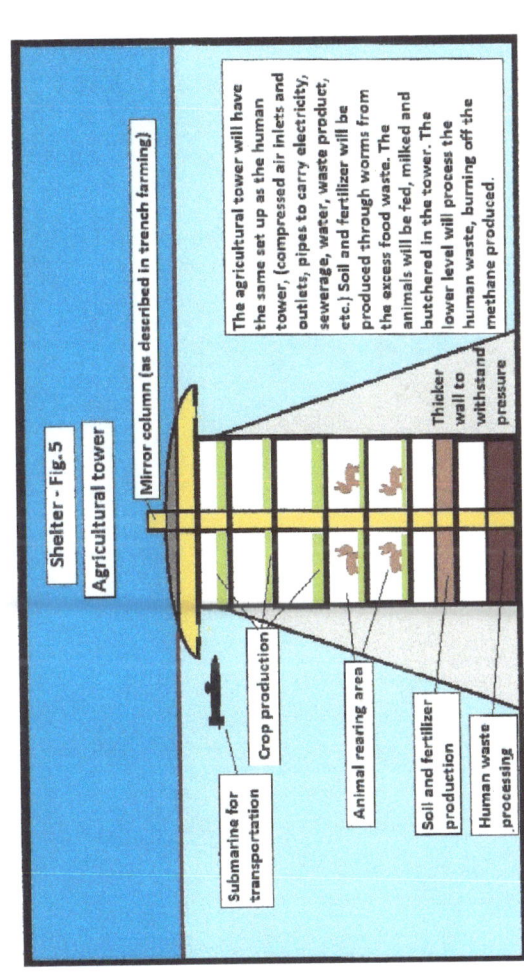

Suffer the Little Children

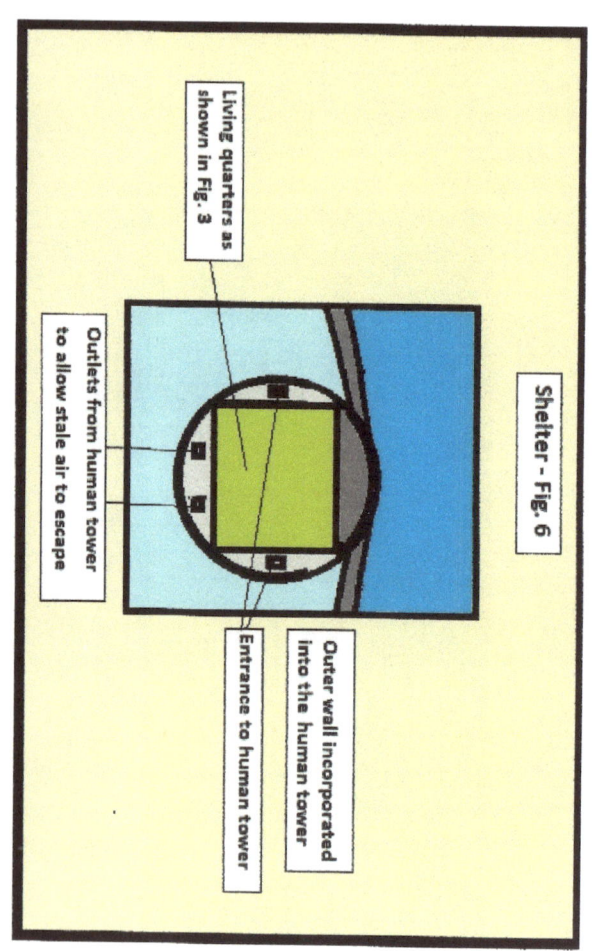

Suffer the Little Children

Electrical Power

One area of concern in our present world is that with the growth of the population there is an ever-increasing demand for electricity. Whereas historically mankind has passed through many stages of development, from the use of wind to drive windmills, to water-wheels, to the development of steam during the industrial revolution and now to electricity, there has always been a price to be paid for such practices. At first the use of natural elements such as wind were harmless, but the problem arose that if a windmill was not in a windy place, its function was limited. Thus, villages sprung up around those areas where the wind blew so that the farmer's grain could be ground to become flour, thus giving a product for market. With the introduction of the water-wheel, the miller was now not dependent on the wind patterns and sought areas where natural waterfalls or weirs appeared. In Europe the downside to water power was that it worked effectively in the warmer climates but with the cold of winter, such rivers could freeze over and thus effect the operation of the flow of water. With the dawn of the industrial revolution a more reliable source of power was needed and with the ingenuity of many inventors we entered the age of steam. Britain led the way in this field but was followed by many European nations and the demand for fossil fuel began, firstly with wood but quickly replaced by coal, which seemed to be so abundant. It was not long before the chimney stacks of factories were spewing out large clouds of coal dust and soot, which began to choke the local environment. Cities that had been used to house only a few thousand people were suddenly inundated with farm hands leaving the fields and flooding into the cities to work in the various factories that were springing up every where. The industrial revolution is still going on today all over the world and as new economies come into existence the demand for coal continues. However, mankind started to search for other means of creating steam, not only for factories but even transportation. Besides the train systems, the first attempts for creating motor cars included ones driven by steam and electric batteries. Oil then came on the scene as an alternative to steam power and

Suffer the Little Children

with the introduction of the internal combustion engine the motor car of today was born. Science then introduced man to a resource that seemed unending and that was nuclear power, using the earth's uranium to power electrical generation plants. As time marches on mankind again is realizing that using the earth's fossil fuels or the atomic power of uranium also comes with a price. The major price has been the pollution of our planet affecting everything from the air we breathe to the water we drink. Whole species of creatures have either gone extinct or are on the brink of extinction. Yet the need for electricity as the major source of power on our planet continues to be created as the world's population increase and countries once deemed "third world" now enter the race for modernization.

When we look at the way man has tried to use the resources of the earth in times past they show us that the evolution of power shown above was not the only option. As early as the time of the Egyptian Empire of the pharaoh's, there is evidence that the knowledge of making dams to hold back water was known and used. Although the dam at Sadd el-Kafara was never completed, probably due to a flood, all the elements of a dam were in place, except for a spill-way, which may have saved the dam. Damming water, either for storage, irrigation or more recently hydro-electric power, is therefore not new but the inherent problem with dams is that in many cases large tracts of land may need to be flooded initially, changing the landscape, destroying a possible ecosystem of plants and animals, causing inhabitants of the area to move elsewhere and creating a possible disaster for persons living down stream, as evidenced in June of 1976, when the Teton Dam in Idaho, USA, broke killing both people and cattle. Another problem is that silt, which normally flows through rivers, is caused to build up at the inner base of the dam, until it has to be removed as it can obstruct or even stop the operation of the dam. However, successful large dams, such as the Hoover Dam, in the Black Canyon of the Colorado river, on the borders of the US states of Nevada and Arizona, have been built to provide electrical power to several states in the area, including California. The main advantage of a dam is that it is environmentally friendly when operating and therefore is a step in the right direction in that regard. Although there is a lack of

Suffer the Little Children

places to build large dams worldwide, the largest dam to date is the Three Gorges dam in China that spans the Yangtze river.

Water power has been used in various ways in history from the primitive water wheel of Egypt, India or Greece, which have been developed over the years into the turbines that we see today. One such turbine, the Francis turbine, is used extensively in hydro electric projects as it has fins that allow the flow of water to be controlled, which is critical for generating the 60 cycles per second in the American continent and 50 cps in Europe. Going along with the use of water wheels and turbines is the use of high pressure water systems, first used to any extent in England during the industrial revolution of the 1800's, but to create such pressure it took the production of steam, and thus the use of coal and coke, to generate such power.

One other power source that came out of this time period, was compressed air. It too had early beginnings as it was used by the smiths of Catalan, Spain in and around 1560. They wanted to use compressed air for their forge and did not want to utilize the traditional hand operated bellows, in use at the time. Someone had the brilliant idea of using water running down a pipe that collected air bubbles as it went. They created a cavern at the bottom of the pipe and found that the compressed air collected was sufficient to power their forge. The process became known as the Catalan forge or Trompe but compressed air was ignored for many years as a power source until a tunnel was to be built in Switzerland. It was 1861 and the project was known as the Mont Cenis Tunnel. Work had been slow using traditional drilling methods of the time, which were mainly manual. It was decided to use compressed air, but it was not known if the distance of seven miles would create a problem. To the engineers' delight the 87-psi used, not only travelled the distance easily but it was found there was little or no change in the psi from the compressor to the pneumatic drills that were used.

In Vienna, Austria there were three native born watch/clock makers named Viktor Popp, Ernst Resch and Albert Mayrhofer who patented a "pneumatic clock". Viktor Popp then made a proposal to the 1878 World's Fair to be held in Paris, France that he demonstrate the clock powered by compressed air. He did

Suffer the Little Children

show his clock (see illustration) and the next year (1879) moved to France starting the "Companie Générales des Horlogues Pneumatiques Système Popp-Resch" with the now Victor, (Victor with a "c"), Popp as its director. Victor set up about 8,000 clocks throughout Paris and by an ingenious method managed to have them keep accurate time by using pulses of compressed air. He also found that there was a demand for compressed air to drive other machinery in Paris and with the city's monetary help constructed a central station at Rue St. Fargeau (Belleville) which used steam driven compressors to supply about 80 psi to a pipeline stretching thirty miles around the Paris area using the local sewers to carry the pipes.

Victor Popp
Pneumatic
Clock

We have an English professor, Alexander B. W. Kennedy, FRS. to thank for giving us a detailed account of the system which he studied in much detail on his visits to Paris, using his experiments to ascertain the efficiency of the Popp system. (Taken from his book - "Compressed Air - Experiments upon the Transmission of Power by Compressed Air in Paris (Popp's System)" dated 1892)

In his own words he states:

"In our own country the distribution and use of high-pressure water has been carried out with the greatest engineering skill and with correspondingly great success by the Hydraulic Power Company in London, and also in Hull.

Suffer the Little Children

Electric transmission (for traction at least) has been at work on a small scale for some time in various places, and is about to be tried on a much larger scale on the Southwark Subway. In America it has been very widely used for traction on tramways, and on the Continent, it has also been used to a certain extent for power transmission for general purposes. Steam has been used on a large scale in New York, but as of yet its success does not seem to be unquestionable. Compressed air has, of course, been used over and over again in rough and uneconomical fashion in connection with tunnelling and boring work, but I think only two practical attempts have been made to utilize it economically and on a large scale for industrial purposes. Of these two, one has been made in Birmingham (*England*) and the other in Paris (*France*). The Birmingham Compressed Air Power Company has established works on a very large scale, but various causes have unfortunately combined to cause delay in the commencement of its operations, which indeed are hardly yet fairly started. In Paris, however, the transmission of power by compressed air has been in operation on a somewhat large scale and with very great mechanical success for a few years past. I have recently had occasion to spend some weeks in making experiments in connection with the Paris compressed-air plant, and have been given the fullest permission to publish the results of my experiments."

"The work now carried on by the Paris Compressed Air Company has developed from very small beginnings, at first slowly, lately very fast. It originated in a pneumatic clock system, which was started about 1870 with a small "central station" in the Rue St. Anne, in the centre of Paris. This business grew gradually until it became far too large to be carried on from such a position, and a few years since a central station, with much enlarged machinery, was established in the Rue St. Fargeau, which is in Belleville, about 4 miles east of the Madeleine. There are now about 8000 pneumatic clocks, public and private, in Paris, driven from St. Fargeau, and regulated by a standard clock in the Rue St. Anne; but as this part of the work, although it originally formed the basis of the whole system, is now a comparatively very small part of it, and is of an entirely special nature, I do not propose to say anything further

Suffer the Little Children

about it here. Until about two years since, a pair of single-cylinder horizontal engines by Farcot, and a beam engine by Casse, sufficed for the whole work; but by that time the demand for compressed air for working motors had so increased that extension had become imperative, and the present working plant of six compound condensing engines, each working two air-compressors, with the necessary complement of boilers, was put down. This plant, except the compressors, was supplied from England by Messrs. Davey, Paxman & Co. of Colchester. The compressors for the English engines were made in Switzerland on the Blanched system. The demand for power is at present so great that, at certain hours of the day, practically the whole plant, old and new, indicating considerably over 2000 horsepower, is fully at work, and in consequence a duplicate main is being laid throughout, and new engines and compressors, half of them constructed by Davey, Paxman & Co., and half by John Cockerill & Co. of Seraing, are being pushed forward as rapidly as possible."

"Summarizing now the whole matter as regards efficiency, it may be said that the result of my detailed investigations is to show that the compressed-air transmission system in Paris is now being carried on in a large commercial scale in such a fashion that a small motor four miles away from the central station can indicate in round numbers 10 horse-power for 20 indicated horse-power at the station itself, allowing for the value of the coke used in heating the air, or for 25 indicated horse-power if the air be not heated at all. Larger motors than the one I tested (and there are a number of such in Paris) may work somewhat more, and smaller motors somewhat less, economically. The small rotary motors would, of course, be much less economical. The figures which I have given are, however, such as can be reached by any motor of between 5 and 25 indicated horse-power if worked at a fair power for its size. M. Victor Popp himself, and the engineers of the company, by no means content with the results already obtained, are experimenting in various directions with a view to greater economy, and I have not the least doubt that they will attain their end. But although I made several experiments on new apparatus, I prefer to leave their results here undiscussed, confining myself as strictly as possible to the work which has been

Suffer the Little Children

already carried out and the economy of actual present working, rather than giving any credit for the result of improvements which, however certain they may be thought, are not yet actually carried out in practical work. A system of transmission which has actually been carried out on a large commercial scale in such a way as to have an indicated efficiency of 50 per cent between prime mover and secondary motor, four or five miles apart, is one which needs no adventitious aids to commend it to notice, especially where its uses are so numerous and so varied, and its convenience so extremely great, as are those of compressed air. Both M. Victor Popp, who has organized and carried through the work of the Paris company, and Mr. James Paxman, who has designed and made the greater part of the machinery used, are to be heartily congratulated on the results which have attended their work."

"While, however, I am unwilling to lay stress on possibilities which are not yet actualities, there can be no harm in saying that I have no doubt whatever that with mere improvement of existing methods and appliances, and without the adoption of any new or untried methods whatever, the new plant of the Paris company now being constructed can be made to have an indicated efficiency of 67 per cent instead of 50 per cent, and a system of transmission which has actually been carried out on a large commercial scale in such a way as to have and to give about 0.54 effective horse-power at the motor for each indicated horse-power at the central station, in the case of such a motor as that on which I experimented. Under these circumstances the air used per indicated horse-power at the motor would be 520 cubic feet, or 650 cubic feet per brake horse-power. I have the less hesitation in giving these hypothetical figures because the more important imperfections of M. Popp's transmission system arise from such a very obvious cause. Nothing, indeed, can be easier than to point out various weak points in the arrangements adopted, and yet, in spite of all such somewhat cheap criticism, the fact remains that no one has yet carried out a compressed-air transmission with anything approaching to the same success on anything like the same scale. The fact is that the success of the system has been essentially due rather to the practical good sense with which the work has been carried through than to

Suffer the Little Children

any special novelty in the methods employed. The air-compressing arrangements at St. Fargeau are in no respect novel or especially perfect; they had been used over and over again before; there is no special advantage in M. Popp's rotary motor that may not probably be possessed by many other rotary motors; the larger motors are simply good ordinary steam-engines, such as can be bought any day in open market, used without the slightest alteration. Of the fan-meter it can only be said that it works well enough to allow progress to be made while it is being improved, and even of the coke stove one would not like to say very much more. The plan of heating compressed air before using it in a motor was first proposed many years ago. The great success which has attended the work in Paris has been attained because its directors have wisely chosen rather to set to work with imperfect apparatus, if only it were simple, fairly effective, and ready to hand, than to wait for the possible invention of novelties and improvements, or to risk the success of their start by the use of any unknown or untried apparatus, however promising its nature. They have had, moreover, the great advantage hitherto of being always asked for more air than they could supply, so that their works have grown and increased simply to meet a growing and increasing demand, and (fortunately perhaps) the urgency of the demand has left them no alternative but to meet it by the very simplest possible means."

Professor Kennedy goes on to tell what kind of industries the compressed air was used for:

"Most of the compressed air in Paris is used for driving motors, but the work done by these is of the most varied kind. A list which I have gives the locality, use, and power of 225 installations, nearly all motors working at from 1 horse-power to 50 horse-power, all driven from St. Fargeau, and the great majority of them more than two miles away from it. In a number of cases the motor drives dynamo machines for electric lighting. In the offices of the Figaro and Petit Journal large motors are also used for printing, and there are many small printing establishments also worked by compressed air. Among the smaller industrial purposes for which the air motors are used in Paris, I find the driving of

Suffer the Little Children

lathes for metal and wood, of circular saws, shearing-machines, drills, polishing-machines, and many others. They are used also in the workshops of carpenters, joiners and cabinet-makers, of smiths, of umbrella-makers, of collar-makers, of bookbinders, and naturally in a great many places where sewing-machines are used, both by dressmakers, tailors, and shoemakers and from the smallest to the largest scale. They find application also in all sorts of industrial work, with confectioners, coffee-roasters, color-grinders, billiard-ball makers, in many departments of textile industry, and other matters too numerous to mention.

I may mention one particular instance of variety of application which interested me much. At the Montagues Russes I found a large horizontal engine placed in a recess driving a dynamo and cells for the electric lighting of the whole building; a small vertical engine in another place worked the rotary pump, which actuated the "cascade"; two or three large air-driven fans in wooden shafts served for ventilation ; and lastly, simple connection on a flexible pipe threw the air-pressure into the beer-barrels as they were brought in, and transferred their contents to a height from which they could afterwards descend by gravity to the place where they were required."

From the statements above, we can ascertain that there were many homes, stores and factories reliant upon the use of compressed air, but of interest is that the pipelines to carry the air spanned an area of thirty miles across Paris and there was very little difference in the air pressure from the Central Station to the different locations considering the many leaks along the way. Victor Popp used cast iron pipes butted together with indian-rubber abutments to achieve his method of supplying compressed air at around 80 psi to his clients. He later used the River Seine as his source of water and built a larger air compressor using a coal powered steam engine arrangement. However, what we need to remember is that compressed air can travel long distances with minimum loss of power and that compressed air can be used to power machinery. (Note that Victor Popp introduced a small heater to expand the air before it entered the machines.)

Suffer the Little Children

Seeing that the above feat was accomplished in 1888 and was adopted as a source of power in Paris until 1994, the use of compressed air is quite significant. Indeed, many factories today, in all parts of the world, carry large containers of compressed air to power many pneumatic machines.

Remembering the discovery of producing compressed air in Catalan, Spain, we mentioned that the apparatus was dubbed a "Trompe". In modern times the trompe has been used to clean rivers where large amounts of iron have been deposited, but a very interesting application has been the production of compressed air using a trompe for pneumatic drills and cooling fresh air for mines. One genius in this field was Charles Havelock Taylor. He was born in 1859 in New Brunswick, Canada and had a father who constructed saw-mills, and travelled from one place to another setting up the mills and taking his family with him. Charles therefore, did not receive much formal education and so he became self-educated and helped his father a lot but as he grew his analytical mind was always looking for opportunities. One such came to him in 1895 when he was working on the construction of a dam in Buckingham, Quebec. He noticed that during the winter the water that ran over the spillway contained a lot of air bubbles that he followed downstream and found that there were ice domes formed. Taking a hammer to a few he discovered that they contained air under pressure and rightly attributed it to the water flowing over the spillway. Although, as far as we know, Charles had not heard of a trompe he could see the potential in capturing compressed air from the flow of water that he had just observed.

He started to make models to both understand the principle of using falling water to make compressed air and then, with the aid of investors, built a small compressor for a site in Magog, Quebec which managed to produce compressed air at 52 psi. He then travelled to British Columbia, Canada to build another one but as the mine did not open as planned the compressor was not used. He then travelled to Cobalt, Ontario, Canada where he designed a compressor that would produce compressed air at 125 psi. When he tried to get investors for his ideas and they heard that his system would have no moving parts and could operate indefinitely so long as the Montreal river was flowing, it was not an easy sell, but he did manage to raise enough funds to build his compressor and to everyone's surprise it worked

Suffer the Little Children

perfectly; until the compressor was taken over by the Ontario Hydro company and closed in the 1980's.

Impressed by the compressor at Ragged Chutes, (the one in Cobalt, Ontario), a Captain Hooper, who was hired to re-open a mine at the North of Lake Superior, in Michigan, USA. thought of using such a compressor for the project. The mine was called the Victoria mine and was close to the West Branch of the Ontonagon river where it dropped 72 feet over the Glen Falls. Charles Taylor met Captain Hooper and found the area ideal for building a Taylor compressor. Part of the river had been dammed so that a fore bay was created that became the inlet for the water feeding the three five-foot diameter pipes, lined with cement, that dropped 342 feet below the ground. The water then met a steel cone that separated the air from the water and allowed the air bubbles to fill a cavern, 281 feet long, 18 feet wide and 20 feet high, along with the water. As the air became compressed it pressed onto the water below but was able to escape by a pipe 12 inches in diameter at a pressure of 117 psi., at the other end of the cavern. If the water level was pushed below a certain point in the cavern it would expose a pressure relief pipe that would allow the air and water to shoot up this pipe and into the air above ground, sometimes creating a geyser effect of water being propelled over a hundred feet into the air. (See Fig. 1 for a simple diagram of the compressor's workings.) The Victoria Mine compressor ran until 1991 and was shut down because a new dam was built that did not feed the fore bay.

Of interest, there is a project to resurrect the Taylor compressor concept in Sudbury, Ontario, Canada but instead of relying on a dam to supply the water, it is to have a pump, (drawing .7 MW of electrical power), to create a circulating action to the top of the trompe, which although seems an excessive use of hydro power, when compared with the alternative; a modern mechanical compressor that draws 1.2 MW of electricity to produce the same amount of compressed air at 125 psi., the savings are substantial.

What we therefore learn from the Paris project and the Taylor compressor is that substantial amounts of high pressure compressed air are possible and that it can be piped for long distances with a negligible amount of loss of pressure.

Suffer the Little Children

Therefore, it seems feasible to suggest that an electrical system could be created that would only use compressed air and water to operate, without the use of any fossil fuels.

I now give some of my ideas to create such a system that can be adapted to anywhere in the world and if not, it would supplement the present system and may also take over from any system reliant on fossil fuels.

The compressor proposed is shown in Fig. 2, which means that this unit can be situated anywhere, where there is access to an ocean, sea, lake, river or even a stream. In areas where the climate has winters that freeze water, the unit will be built either underground or else solar heat may be used to keep the water flowing. If the area is subject to high degrees of evaporation the unit may again be placed under ground or an umbrella effect will be created to stop such evaporation. Either way the unit is versatile enough to be used in any climatic situation. (See also Fig. 2a, 2b and 2c.)

Lessons learned from both the Taylor compressor and Victor Popp's system would be used to help the project work.

Fig. 3 shows a tower built 80 feet high with a diameter of 200 feet. 60 feet of the structure will be above the water line and 20 feet will be below. Central to the tower will be a storage tank that will be forty feet in diameter and twenty feet deep. This will act like a battery and store the energy to run the system. It will be fed by various methods that will be described later. The water from this storage tank will be siphoned into the venturii cones of two sets of trompes. One trompe will go to a depth of 75 feet and produce compressed air of 42 psi. while the other trompe will go to a depth of 150 feet and produce compressed air at a pressure of 84 psi. The compressed air from the second trompe will be piped to a water pump reservoir that will increase the air pressure from between 120 psi. to 350 psi. depending on the use and distance the air will need to travel.

In the tower will be sixteen such sets of trompes, as shown in Fig. 4, which are positioned between pie shaped pools of water twenty feet deep. These pools are fed by the outer water, which will be filtered as it is siphoned into the pool. It is these pools that

Suffer the Little Children

allow the central storage tank to keep filled, and thus it is important that there be more than one way to do this.

Fig. 5 and Fig. 5a show the workings of the Savonius windmills that will top the dome of the tower. Each windmill is divided into sections, where circular permanent magnets are placed in sets of four, showing a North then a South face. The permanent magnets are arranged on a spindle so that the movement of the windmill vanes causes the magnets to spin. Sandwiched between them are electromagnets that will be fixed, and their leads will be attached to a common feed wire, thus the electric current generated by the movement of the permanent magnets will be alternating and in series. The common feed wires will pass down through the dome above the pie-shaped pool. Wires will then come from the feed wires and be wound around stainless-steel pipes. The windings will go firstly in a clockwise direction and then after the coil is about ½ inch thick the windings will be reversed for another ½ inch. The purpose for this is to create an effect first discovered by Nikola Tesla. He wound a flat coil of insulated copper wire and then wound another flat coil in the opposite direction, which meant that when alternating current passed through the coils, an induction between them was set up. (This is the basis for the modern induction cookers starting to make their way into our kitchens.)

The induction process will then be started by the action of the Savonius windmills and thus the stainless-steel pipes will begin to be heated. The pipes will be immersed in the pie-shaped pool so that the water within the tubes will begin to boil. The steam thus created will pass up the pipes and into the central storage tank, where it will condense and turn back into water. Thus, the induction pipes have created a pump to pass the water in the pie-shaped pools up to the storage tank, the pump pushing the water about one hundred feet high.

The draw-back with this system is that although the electricity will be generated while the Savonius windmill is spinning, if the wind drops there will be no power to the induction pipes. To compensate for this the first trompe that descends 75 feet will produce compressed air at about 42 psi. This will be fed into the system I call a gravitational pump, shown in Fig. 6. The principle it works on is that if compressed air of a low pressure is pumped into a closed container, it will continue to compress. Thus, if that

Suffer the Little Children

container was filled with a liquid such as water and the compressed air was pumped into the water, the compressed air would collect at the top of the container. If a weighted plunger is wedged into a hole at the top of the container the air pressure would continue to increase until there was enough pressure to force the plunger out of the hole. The gravitational pump uses this idea but instead of one plunger there are several plungers of different weights. Therefore, when the air pressure in chamber one moves the plunger one, the compressed air entering chamber two will be higher than it was in chamber one. Chamber three then has a plunger that is heavier than chamber two and again the air entering will be compressed further.

It is envisaged that there will be enough stages in the gravitational pump to take the input 42 psi. to a compression of 120 psi. in the output of the final chamber. This compressed air, leaving the final chamber, will then enter a pipe, which will be in the surrounding water, and as the air enters this pipe it will create an air-lift pump which will carry the water in the pipe to a height of 100 feet allowing the water to also enter the storage tank. However, the compressed air will not be a continuous flow but will be intermittent bursts as the pressure builds in each chamber. This will be fine because an air-lift pump acts best when air bubbles push the liquid up the pipe and so bursts of compressed air will work for our purposes.

The gravitational pump will then be our second method of raising water into the storage tank and will act even when water is being pumped by the action of the Savonius windmills. There will be sensors in the storage tank to alert either a computer and/or manual workers that the storage tank is either too full or too low. Should the Savonius windmills stop turning or if they are not producing enough electricity to power the induction pump sufficiently, then there will be tanks of compressed air filled by the second trompe of 150 feet, at 84 psi. This air will be directed at more pipes in the pie-shaped pool to provide another air-lift pump system.

Fig. 7 shows the second trompe and its exit to the tunnel that leads to the compressed air storage on shore.

Each on-shore storage silo will be fed by four of the sixteen trompes in the tower, which will raise the top half of the unit, (Fig.

Suffer the Little Children

8 and Fig. 8a), which will make it available for the water pump action to begin. The silo compressed air pump is similar to the water pump, but instead of having floats, the compressed air from the previous silo will raise the pump with its empty tank on top. When the water from the top tank overflow is siphoned into this tank it will act like the water pump and push the compressed air in the unit through the venturii funnel, which in turn will further compress the already compressed air. The result will be a sizeable air lift pump which will push the water in the overflow tank below up to the upper tank. (Note that the compressed air will escape at the top, but the water will fall back into the upper tank.) A spillway will take the overflow to allow the siphon, that will be controlled by a valve, to enter the empty tank of the compressed air pump. Any excess water will return to the overflow tank below. The Tesla turbine arrangement will then turn conventional electric generators to supply any needed electricity to power the valves and silo plant. (The main electrical generation plants will work on the same principle but with much more compressed air and other trompes attached.)

The compressed air from the silo will either be used as the air to travel 20 miles to the next station or it will be used to generate electricity for the tower as shown in Fig.8. The electricity from this generation plant may also be used in the tower to power lights, computers and other equipment used by the workers. It will also be a back up resource for the pumps that take water from the pie-shaped pool to the storage tank.

At the end of this section there is a list of different ways to raise water, so that any of such methods may be used either in the tower or the generator. The important factor as far as the tower is concerned, is that the storage tank must be constantly full, thus allowing a head of water for the trompes. As has been described above, to achieve this it may be necessary to only run one set of four trompes or if the wind power, solar power or other alternative source of power fails, the storage tank will keep the operation going until the wind starts blowing again or the sun starts shining.

As mentioned above the compressed air in the on-shore silo will be used to travel to the next station. This next station will be another similar silo set-up with the exception that it may or may not have an electrical generator attached to it. We will assume

Suffer the Little Children

that we wish to connect our compressed air system to a town or village that needs electricity, that is sixty miles from the on-shore silo. Since we are transferring compressed air only twenty miles, it will require two stations plus the original station to transmit the compressed air the sixty miles. When the compressed air reaches the town or village it will enter another station that will be set up with an electrical system, which will then supply electricity to that town or village, (See Fig, 9)

The distribution system will be modelled on that of Victor Popp, of using underground pipes, (remember he used the Paris sewer system), to carry the compressed air. Whereas, Victor Popp used cast iron for his piping we would use stainless steel pipes and as much as possible have them travel in straight lines instead of curving. To accomplish this the pipes would be housed in tunnels that would be placed sixty feet below the surface, thus eliminating most obstacles to a straight course. Such tunnels would be seven feet in height and be shaped like the cross-section of an egg, thus duplicating the powerful strength of an egg-shell. The tunnels would allow workers to inspect the piping for possible weaknesses or leaks so that they can be repaired or replaced before any major problem would arise. Since these tunnels will need air for the workers, there would be vents at certain points along the tunnel, but to avoid flooding these vents would be a structure, such as a lamp as seen on the side of a highway. In this way the entrance for the air can be kept well above the ground.

Besides the major artery to the town there will be other towers built either in lakes, rivers, or be placed inland but with a self-contained lake of water, and these towers will feed the main artery system. Thus, the idea would be to replace the need for power lines that at present clutter our countryside. (Such power lines are more efficient than compressed air for their ability to transfer power with less losses. but power lines are also more susceptible to weather conditions, particularly snow and ice storms, and therefore it is felt that the compressed air pipeline will be more reliable.) What is envisaged is an eventual replacement of the present electrical grid system for a localized electric generation plant that will supply the local needs of a community. This localization should mean that there would be less black outs, that are becoming all too common with the present grid system. It also means that environmentally the

Suffer the Little Children

compressed air system will not cause the pollution that coal-fired electrical generation plants do presently. It will mean that we can tap into the inexhaustible resource of the air around us without using it up, in fact exhausting isothermal air back into the atmosphere will give the air higher quality than we have at present.

Ultimately it may be that towns will spring up around the twenty-mile mark of where the compressed air meets another station. It may be that an industrial plant or factory would be built around the station and thus attract workers to settle in the area; followed by major chain and grocery stores. As the system grows the idea of supplying major cities with electricity becomes a possibility but at the other end of the spectrum, rural areas and impoverished areas of the world may be also be supplied with the compressed air/electric system. In rural areas, local generation plants would not need to have trompes attached as the unit may only supply one farm at a time.

Aquifer style compressed air storage

Up to now we have spoken of silo storage but when the compressed air is to enter an area where the electricity is to be generated it may be necessary to have large amounts available for this operation. There are projects that use abandoned salt caverns to store compressed air and then when it comes to the surface, the air is heated by natural gas and passed through conventional turbines to produce electricity. The aquifer style compressed air storage is along these lines, but instead of looking for natural salt caverns, man-made caverns, much like the one in the Taylor compressor, could be constructed but on a much larger scale. Fig. 10 illustrates one such idea, where a site close to a river is selected, though the river only needs to be flowing and does not need a large head of water, created by a dam. The construction of the aquifer would be made on dry ground and begin by creating a hole, probably about ten feet in diameter and about one hundred feet deep. From this the cavern is created by constructing sections of arches that will support large domes. At the pinnacle of each dome will be a pipe outlet for the compressed air. The pipe outlet may then join other outlets or go directly to the electrical generation plant. Another hole would then be dug at the opposite end of the cavern complex and be ten feet in diameter and one hundred feet deep,

Suffer the Little Children

joining to the cavern. The first hole would act like a trompe although other pipes from various silo storage tanks would enter the cavern at its base. There would also be several relief blow-off pipes coming from the aquifer cavern, similar to the Taylor compressor relief pipe. After the construction is finished there would be channels going to the river, which channels would contain filters to allow only water to pass through them. Such water would then enter the first hole and fill it and the aquifer cavern, exiting via the other hole and back to the river. The pipes from the silo storage tanks would then be opened and the aquifer water level will drop as the compressed air enters, filling the ceiling of each dome with such compressed air. This will mean that vast amounts of compressed air will be available for electrical generation, whether it be by conventional methods or by raising water to create a head of water to run Tesla turbines that in turn will run electric generators. Much of the construction of the aquifer caverns will be done underground without disturbing any land on the surface, and the arches will support not only the domes but the land above the domes.

Clock Compressed Air Storage

Another method of storage, and easier to construct than the aquifer style compressed air storage would be a method I have entitled clock compressed air storage, shown in Fig. 11. Here the compressed air from any source is piped into twelve storage units that are connected to a large pool of distilled water. As the air pushes the water out of the storage unit a float descends a central pole and when the water is at the bottom of the storage unit, the float activates a switch that turns off the compressed air supply to that unit. (Fig. 11a & 11b) The water is therefore kept out of the unit by the pressure of the air. When the valve at the top of the storage unit is opened to allow the compressed air to escape and either power an electrical generator or act as a air-lift pump, the water will return to the storage unit and the float will begin to rise. As the unit empties itself of the compressed air and the water reaches the top of the unit, the float will again trigger a switch to shut off the outlet valve and allow new compressed air to fill the unit. The trigger would also open the next storage unit, which would be filled with compressed air, and the process would continue around the twelve storage units, with the water in the pool rising and falling slightly as it fills each unit. The pool will

Suffer the Little Children

be covered by a dome with a small aperture at the top to allow air to enter but keep evaporation to a minimum. This storage method may be linked to any source of compressed air, whether the silo or aquifer type and allow a constant flow of compressed air. Its advantage will be that the pressure of the water in the pool should compensate for any loss of pressure in the compressed air that may have been the result of it travelling long distances or from leakage along the way.

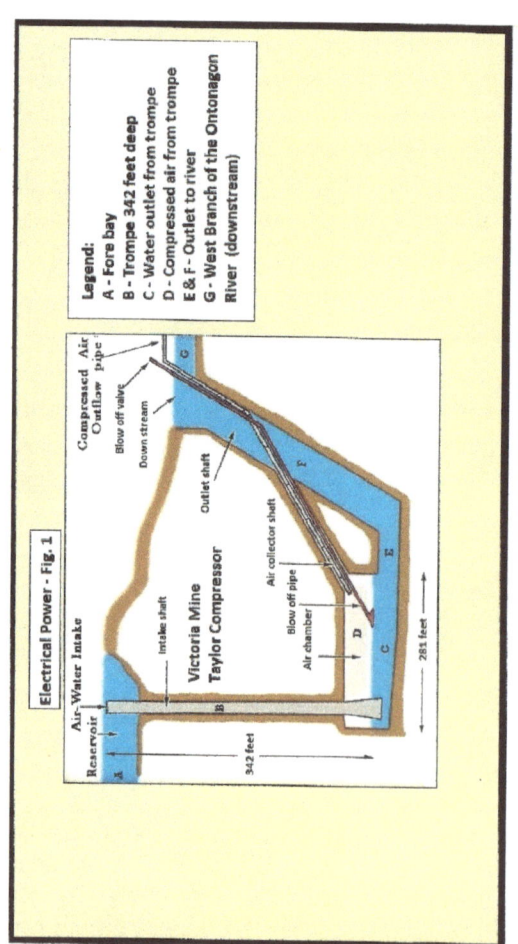

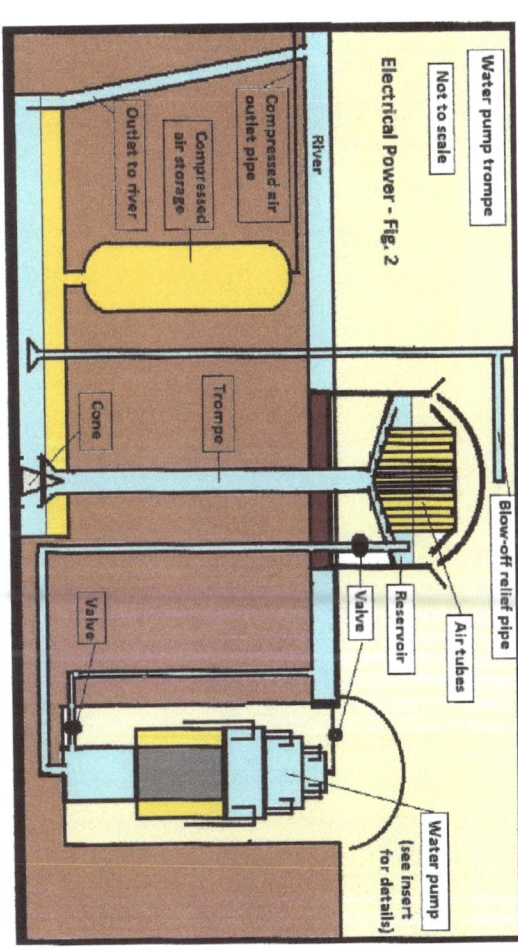

Suffer the Little Children

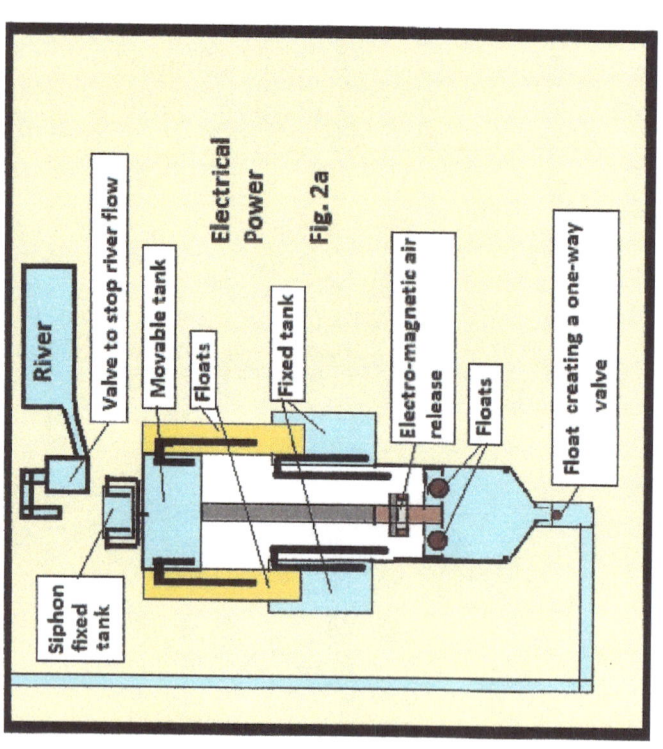

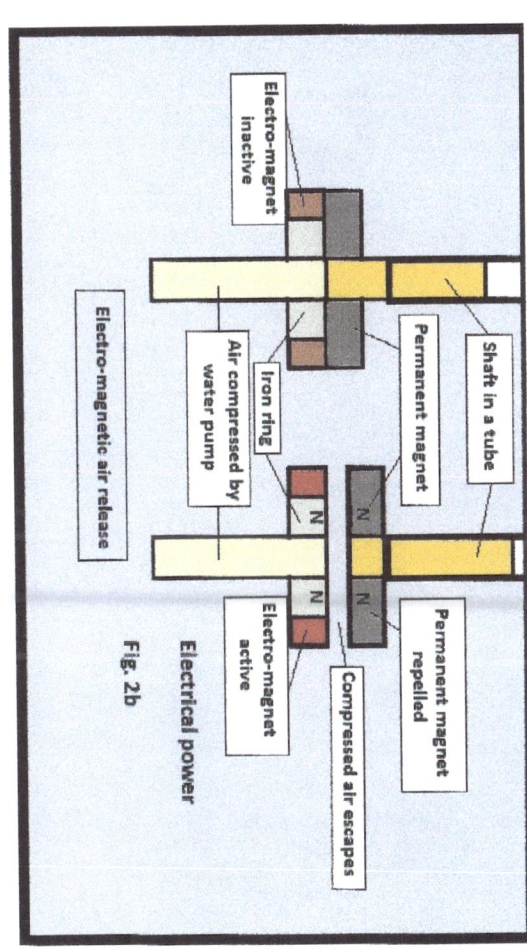

Fig. 2b

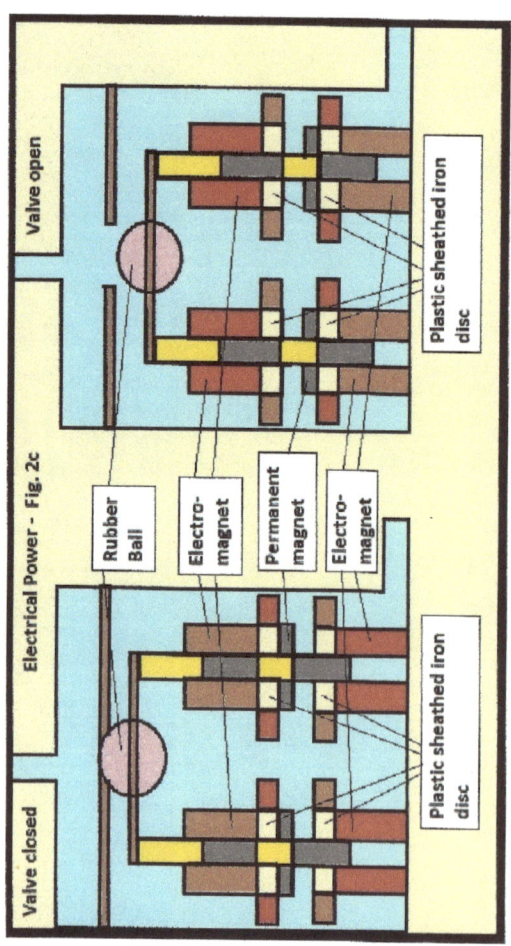

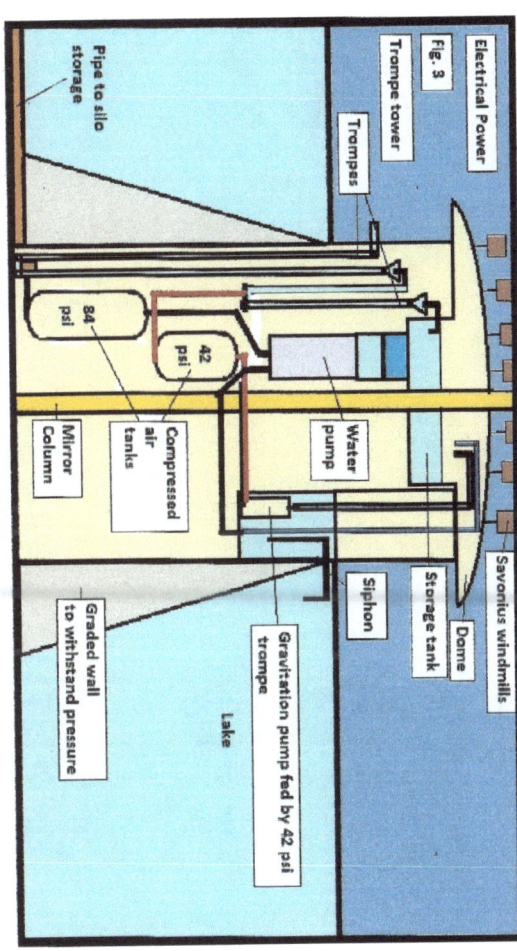

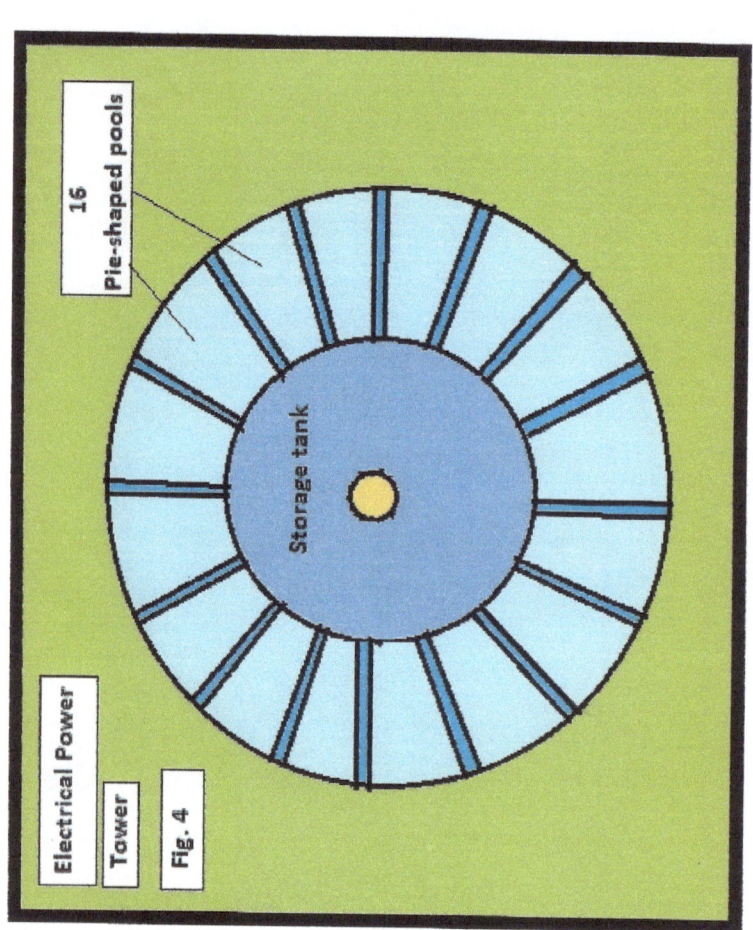

Fig. 4

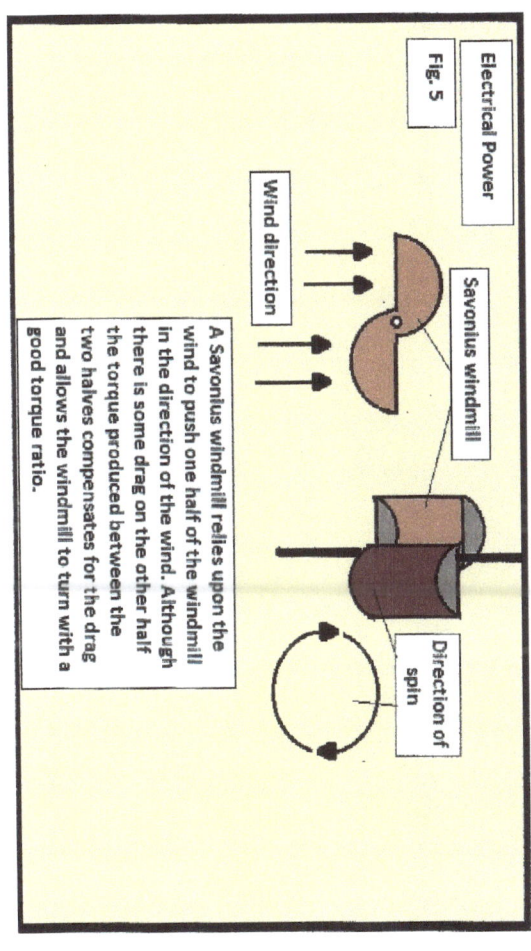

Fig. 5

A Savonius windmill relies upon the wind to push one half of the windmill in the direction of the wind. Although there is some drag on the other half the torque produced between the two halves compensates for the drag and allows the windmill to turn with a good torque ratio.

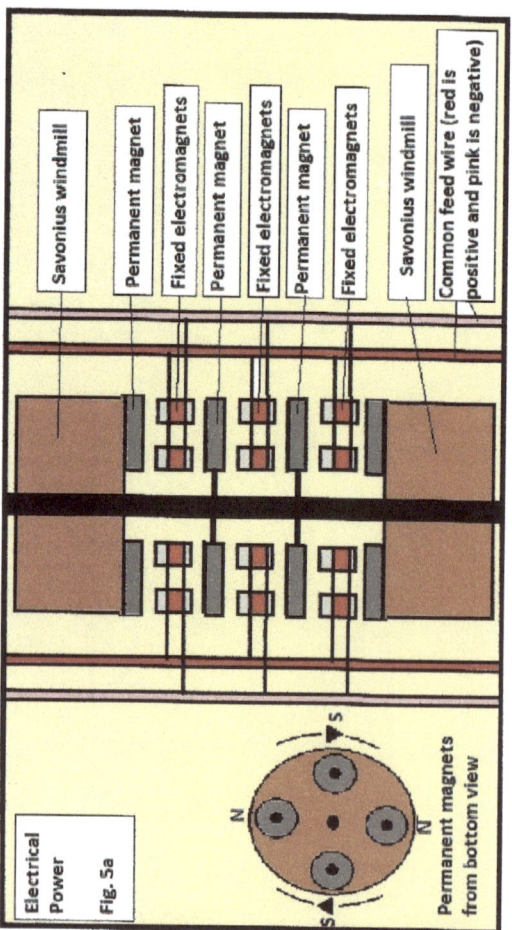

Fig. 5a

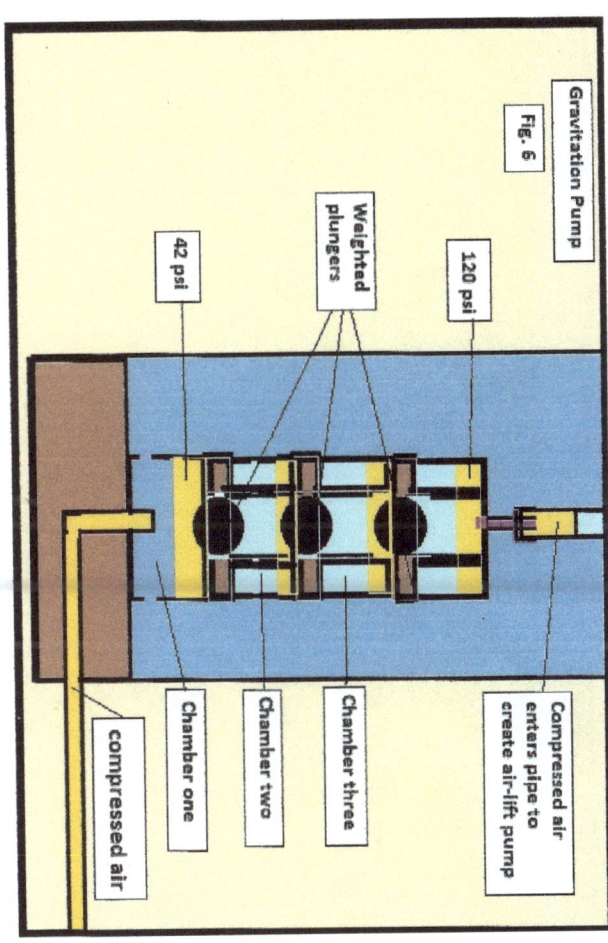

Suffer the Little Children

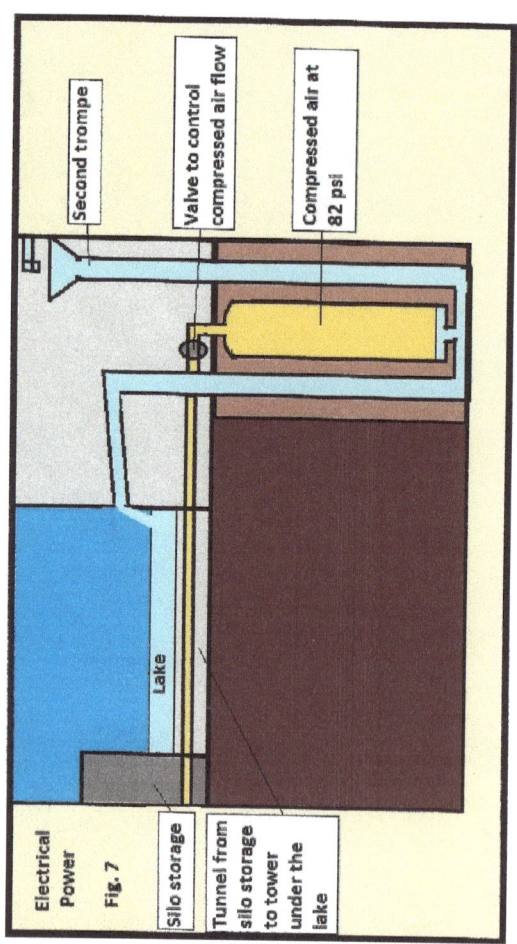

Fig. 7

Suffer the Little Children

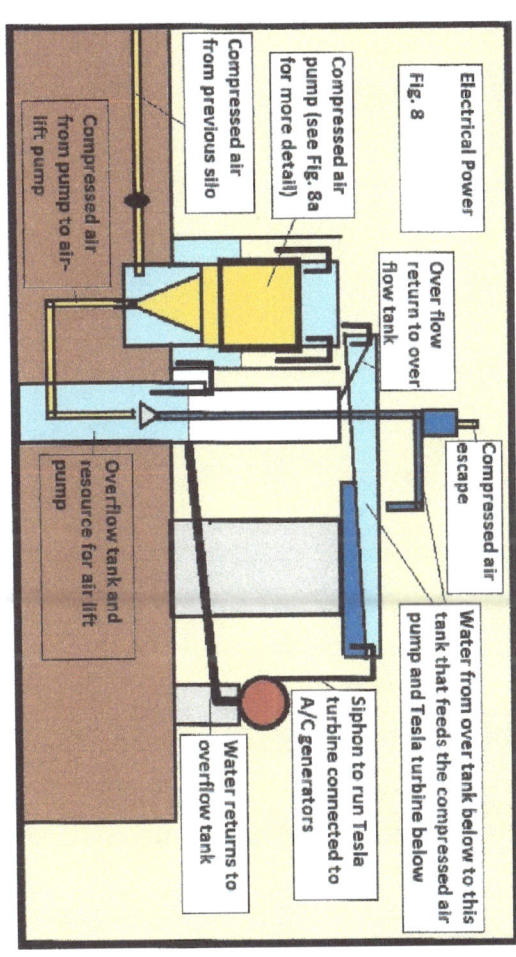

Suffer the Little Children

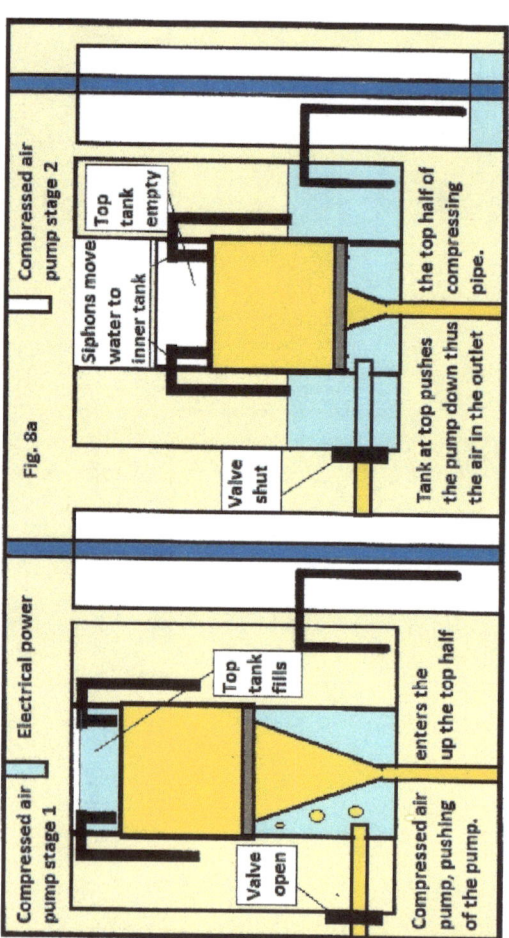

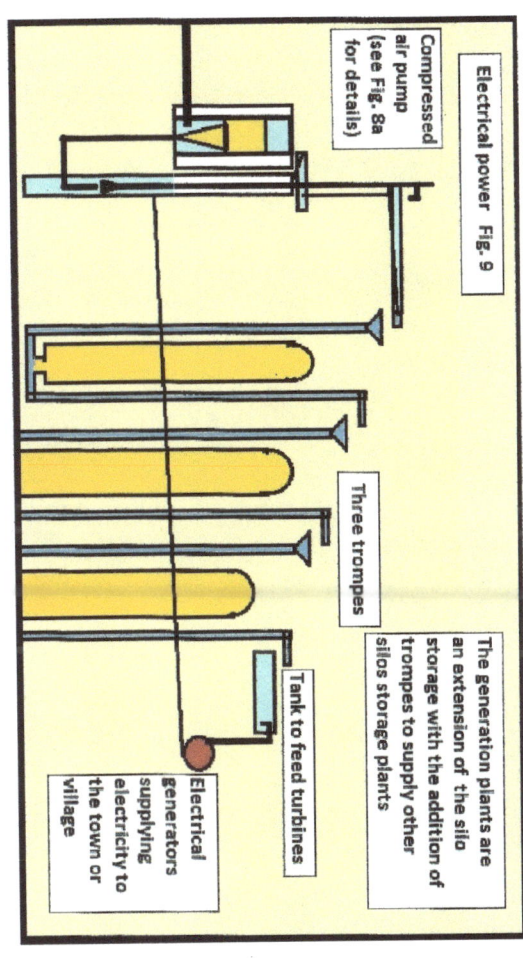

Suffer the Little Children

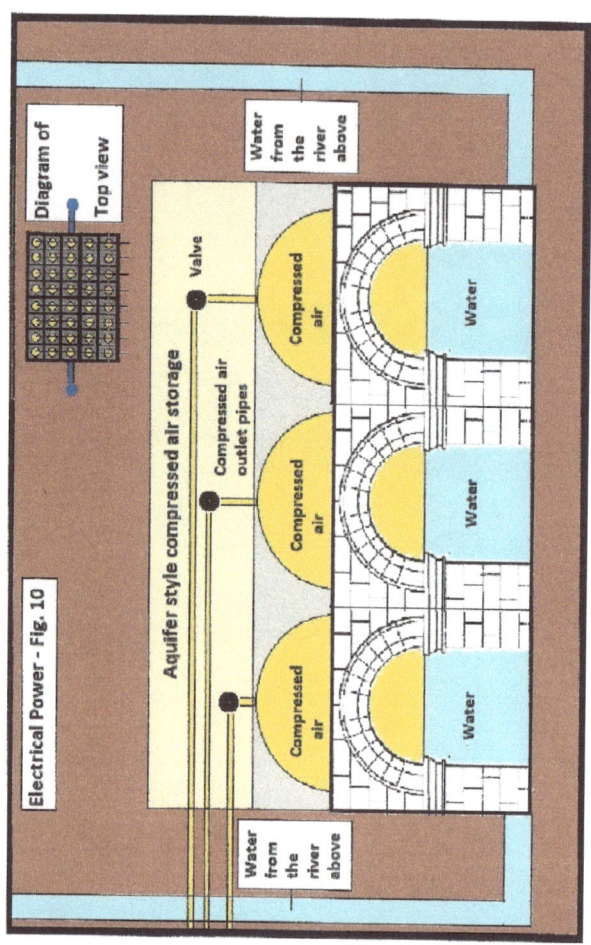

Suffer the Little Children

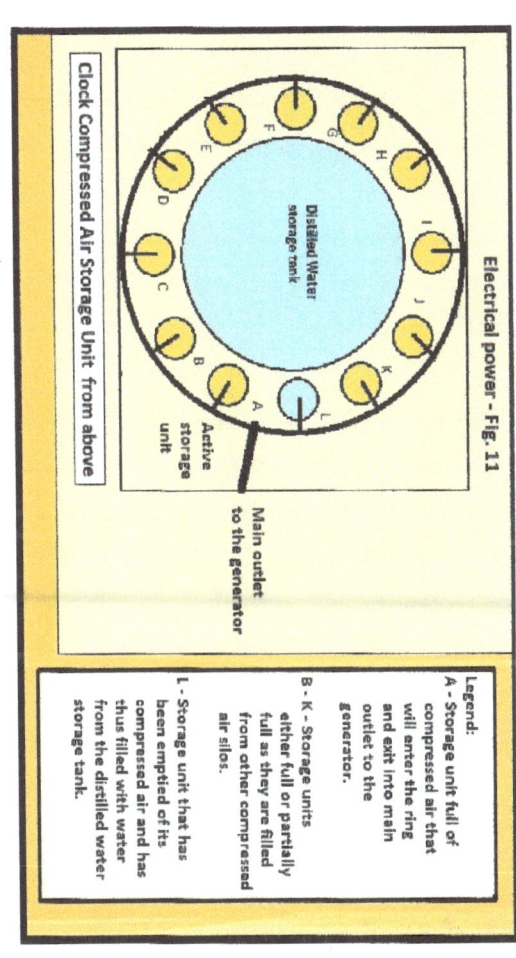

Clock Compressed Air Storage Unit from above

Electrical power - Fig. 11

Legend:
A - Storage unit full of compressed air that will enter the ring and exit into main outlet to the generator.

B - K - Storage units either full or partially full as they are filled from other compressed air silos.

L - Storage unit that has been emptied of its compressed air and has thus filled with water from the distilled water storage tank.

Suffer the Little Children

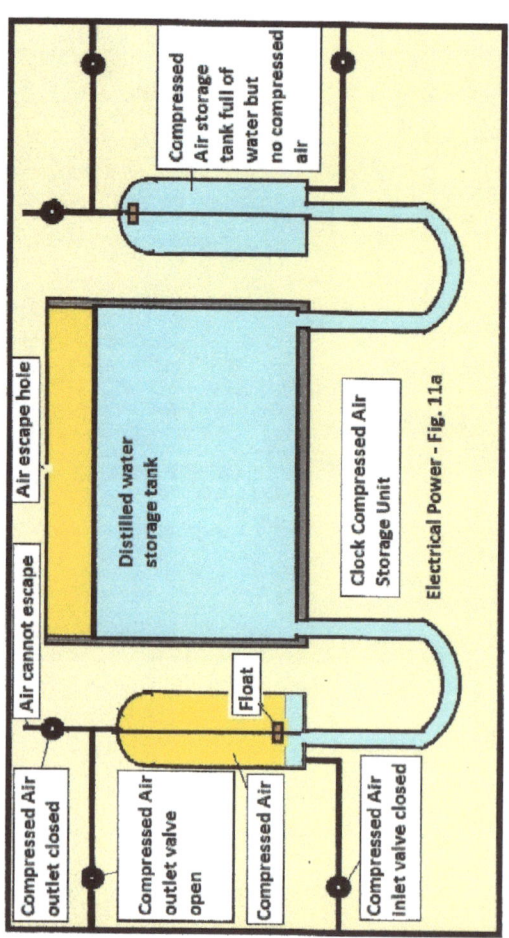

Electrical Power - Fig. 11a

Suffer the Little Children

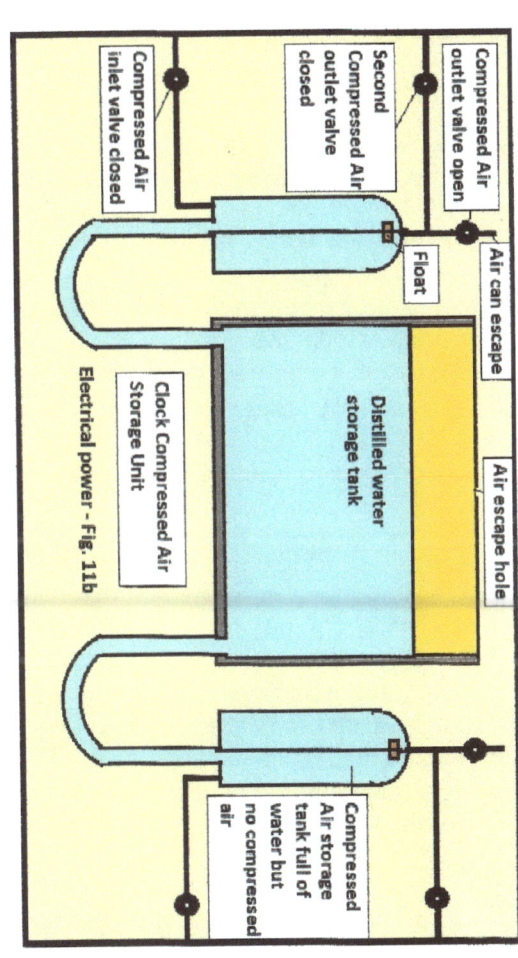

Electrical power - Fig. 11b

Suffer the Little Children

Suffer the Little Children

Methods of Raising Water

In the last section it was mentioned that there is more than one method for raising water to make the air/electric system work. Up to this point we have shown that water can be raised by the water pump, the gravitational (air-lift) pump, the induction steel pipe method and the basic air-lift pump methods. However, I will describe a few more below, including the cavitation cyclone method, the oxy-hydrogen method, the Tesla turbine pump and the pyramid solar furnace method.

When we look at the tower, its main purpose is to raise water high enough to keep the trompes working to produce compressed air. However, we have also explained the compressor, built by Charles Havelock Taylor, relied on the difference in levels created by a dam and a good flow of water coming from a major river. Naturally this same action is created by a waterfall, so it is conceivable that a fore bay would be created at the top of a significant waterfall and thus the outlet would be the bottom of the falls. Such falls worldwide may include Niagara and Victoria but virtually any waterfall of any significance would work. Thus, as Charles Taylor proved, compressed air can be produced with no mechanical means.

Cavitation Cyclone

The cavitation cyclone method relies on a principle discovered in marine engineering. When a boat or submarine has a propeller and it begins to spin in water, bubbles will be created which are liquid-free zones of low pressure, (known as cavitation). These bubbles will then be compressed by the pressure of the water around the propeller, which will cause an implosion of the bubble. In practice this implosion can harm the propeller and thus is not usually desirable, but the implosion also creates heat. It is this heat that we are interested in. Fig. 1 shows a circular block of metal that has many holes bored into its sides. The block is encased in a container that allows a small gap between its walls and the block. There are also two outlet pipes coming from the side of the container. When the block is attached to an

Suffer the Little Children

electric motor and spun quickly, and the container is filled with water, cavitation will begin and in a very short time the heat will cause the water to turn to steam. This steam will then exit via the two outlets and if these outlets are of any length the steam will rise quite a distance. Fig. 2 shows the whole operation and so the cavitation cyclone will now act as a pump turning water into steam that can rise the hundred feet necessary for the tower and its storage tank.

Turning water into steam so that the steam can act as a pump is not new, in fact the reason we have rain is due to the evaporation of water into vapour, and its rise into the atmosphere to create clouds, which eventually falls as rain is a common occurrence. However, it does not seem to be a common answer to pumping water.

Oxy-hydrogen Generator

The oxy-hydrogen method is simply splitting water into its two elements, hydrogen and oxygen and allowing the mixed gas to explode releasing its energy. The method was discovered in 1800 by William Nicholson but he also realized that it takes more energy to split the water into it's elements than it gives off as energy when exploded. This explains why this method is not good for driving a motor of any kind. However, in our project we are looking at the oxy-hydrogen gas as a means of turning water into steam and thus into a pump. Fig. 3 shows how this will work, but the use of oxy-hydrogen gas will be used in the generation of electricity and not for use in the tower. The reason being that to have the process work efficiently there needs to be three basic ingredients. First, we need stainless steel as the electrodes that will split the water into hydrogen and oxygen, the reason being that this metal will not be used up in the process. Secondly the water used must to be distilled as there can not be any foreign elements in the water except for one, which would be number three on our list, potassium hydroxide, (sodium hydroxide may also be used), which acts as a catalyst. (Thus, the process would not work in the tower.)

Direct current would be used to make one stainless steel electrode positive and the other negative. The distilled water with a small amount of potassium hydroxide (KOH), will act as the

83

Suffer the Little Children

electrolyte and the process will begin. The gas is collected in a bubbler, (a reservoir that prevents the gas from exploding back into the main gas generator), and is then passed to a metal tube placed in water. The gas will form bubbles that will come to the top of the metal tube where it will be ignited and turn back into water after releasing its heat. The heat in turn will boil the water both in the tube as well as in the tank so that steam will be created; which will be channeled up to the storage tank of the electrical generation plants. (This method is covered in more detail in another book I have written "Can Tesla Save the World?".)

Tesla Turbine Pump

Another method that is presently used to pump water, is the Tesla turbine. Nikola Tesla dubbed this invention his favorite and is his answer to creating a turbine that will be 90+ percentage efficient, (Tesla mentioned 98 percent.) When Nikola looked at a turbine he saw that it used a lot of waste energy by directing the power in many directions, some pulling and some pushing. He decided to use his knowledge of the Bernoulli findings, that air passing over a curved surface will create a boundary layer that will cling to that surface. This knowledge is critical to understanding how an airplane wing works but Nikola saw it as an efficient method of driving a turbine. He experimented and ended up using ten flat, thin metal discs, each having three holes near to the centre of each disc, (see Fig. 4). By allowing the discs to be arranged close together with a small air gap, Nikola realized that he could produce the Bernoulli boundary layer between each disc which meant that the power source would cling to the surface of the disc and drag it around. He also calculated that a vortex would be created such that the liquid or gas used to drive the turbine would follow a path to the exhaust holes. He used steam to drive a ten- inch diameter turbine which was connected to an electrical generator. Unfortunately, at the time Nikola invented his turbine we had not discovered materials like carbon fiber and so his steel plates buckled when the optimum turbine speed was reached. For this reason, the idea was not adopted although we will use it when generating electricity, however, our purpose for the turbine is its ability to also become a pump. The advantage of such a pump is that

Suffer the Little Children

whereas most conventional pumps clog up with debris easily, the Tesla turbine pump is not hampered by such material.

Fig. 5 shows how the pump works, in that instead of the power source entering via the perimeter of the discs, it enters in the exhaust holes and then the vortex acts in reverse, casting the liquid out at the perimeter. To avoid complicated seals to stop water from entering the electric motor, our pump is positioned horizontally and with the umbrella on the shaft connecting the turbine to the motor, water does not enter the motor. When the water reaches the perimeter the exhaust pipe will have a bend that will allow the water to be forced upwards. Note that there is a venturi effect in the pipe so that it enhances the lift to the water.

The electrical source for this turbine will come from solar panels, situated between the towers as shown in Fig. 6. (Note that in colder climates there will be solar heaters to keep the solar panels functioning during winter.) Whether the direct current from the solar panels will be used as is or passed through an inverter to a battery will come from experimentation, but the motor for the Tesla turbine pump will be powered by the solar panels. By this method the turbine will be just below the surface of the pie-shaped pool the water will be shot upwards into the storage tank.

Pyramid Solar Furnace

There has been a lot of research about using the power of the sun to provide electricity. I have mentioned solar panels, but although they are a method of producing direct current they do not have a high efficiency rating although as time passes their cost is decreasing. Another use of the sun has been to use its thermal energy to heat a liquid using heliostat mirrors, so that an array of mirrors reflects the sun's rays onto a tank atop a tower which contains water. The heat generated turns the water into high pressure steam that is channeled to a conventional turbine and therefore generates electricity.

Another use of thermal power has been to take a pipe filled with a synthetic oil and place it in a parabolic mirror trough. The pipe is painted black and as the sun's rays hit the reflecting surface of the trough the black pipe picks up the heat and the oil within it

Suffer the Little Children

begins to circulate. The pipe then descends into a tank of water and imparts its heat to the water, causing it to boil and run a conventional turbine to generate electricity. The oil then completes its circuit and returns to the bottom of the exposed pipe to be heated again. (One solar furnace in France, called the Great Solar Oven of Odeilo, has reached a temperature of almost 6,500 degrees Fahrenheit.)

My idea is to use a similar arrangement as above but instead of producing steam for a turbine we will produce compressed air. Fig. 7 shows the set up and you will then see why I have called this invention or adaption, a pyramid solar heater as the exposed pipes run up the sides of a pyramid shaped structure. As with the solar furnace above, the trough is highly polished, and the pipe is painted an absorbing black colour. The pipes run almost up to the apex of the pyramid where they enter a long shaft made of concrete and cement. The shafts run down into a pool of water which overspills into a venturi effect funnel that has several stainless-steel pipes imbedded into it. Below this is a shaft, six to nine feet in diameter which descends 351 feet to a stainless-steel cone which then enters four caverns, each at least 281 feet long, 20 feet high and 18 feet wide. The caverns then end and join with the bottom of the pipes in the troughs, in the pyramid, via a canal-type lake. (The measurements are a combination of the Ragged Chutes and the Victoria compressor discussed earlier.) As with the Taylor compressor there will be a blow-off relief pipe, but this will rise to the top of the pyramid where its geyser will hit the inside of the apex and fall back into the pool where the steam pipes are condensing, thus any blow off due to increased pressure will be able to return to the 351-foot shaft.

Fig. 8 and 8a shows the operation of the pyramid solar furnace. The shaft and exposed pipes will be filled with distilled water to the level shown. Note that there are parabolic mirror troughs set up at the base of the pyramid that will carry the synthetic oil, mentioned above. The pipes from this arrangement will enter the pyramid and be the means of boiling the water that is in containers shown in Fig. 9. This boiling water will enter the exposed pipes that extend the height of the pyramid as shown in Fig. 10. They will begin to absorb the sun's heat and will sustain the steam as it rises to just below the apex of the pyramid. The

86

Suffer the Little Children

steam will then enter pipes surrounded by cement that will enter the pool at the pyramid apex, (see Fig. 11), where the steam will condense back to water. The pool will also contain the air tubes for the trompe and the blow-off pipe outlet from the cavern below. The pool will therefore act as a condenser for the steam, a receptible for the excess compressed air and water from the cavern below, and the over-flow from the pool will go to the trompe below, (see Fig. 12). There will also be vents in the apex of the pyramid to release the compressed air from the blow-off, back into the atmosphere.

As mentioned above, the diameter of the trompe pipe will be from six to nine feet and will fall 351 feet to the caverns below. There the water and compressed air will strike a large stainless-steel cone which will help to direct the mixture into one of the four caverns, (see Fig. 13). The same process as described in the Victoria compressor will occur, with the water causing the air to compress further until at least a pressure of 125 psi will be reached. This will be channeled away by four 12-inch exit pipes and go to the silo storage tank, previously described, that can be up to twenty miles away. As mentioned above, the blow-off pipe will work in the same way as the Victoria compressor, so that if the water level goes below the blow off pipe's entrance, excess air and water will shoot up the pipe and into the pool at the apex of the pyramid, where the over-flow will allow it to return to the trompe. Of course, this will also lead to evaporation and thus a method of topping up the pool will be needed to keep the volume of water circulating. One such method would be to dig a well into an aquifer and pump that water into a blackened pipe that would go in a separate shaft and condense into the pool thus compensating for the evaporation of the blow-off. Water may also be used, by siphoning it long distances as described in an earlier chapter.

At night, there will be Savonius windmills to generate compressed air, which will be used as motive power for air-lift pumps on the North side of the pyramid. In this way, the trompe will be able to function when the sun is not shining. (We will assume that the wind is blowing at night.) During the day the sun will shine on the east side of the pyramid then the south side and then the west side as it sets, allowing the north side to work for

Suffer the Little Children

the system at night. It may also be feasible to have towers as previously described to pump more compressed air to the north side and therefore act as air-lift pumps. The water pumped will be that contained in the pyramid canal as this will be constantly recycled, with the addition of some water lost to evaporation.

So, as we can see we have several options to either move water from the pie-shaped pool to the storage tank, or raising water in the generation plant, or using solar energy to create an entirely separate tower system. Obviously. the climate will play a large role in determining which method or methods to use but ultimately the system is feasible regardless of its location in the world. Thus, as we mentioned earlier there is a way to generate electricity without pollution, without harming the environment and without worrying about where the electricity is needed.

Suffer the Little Children

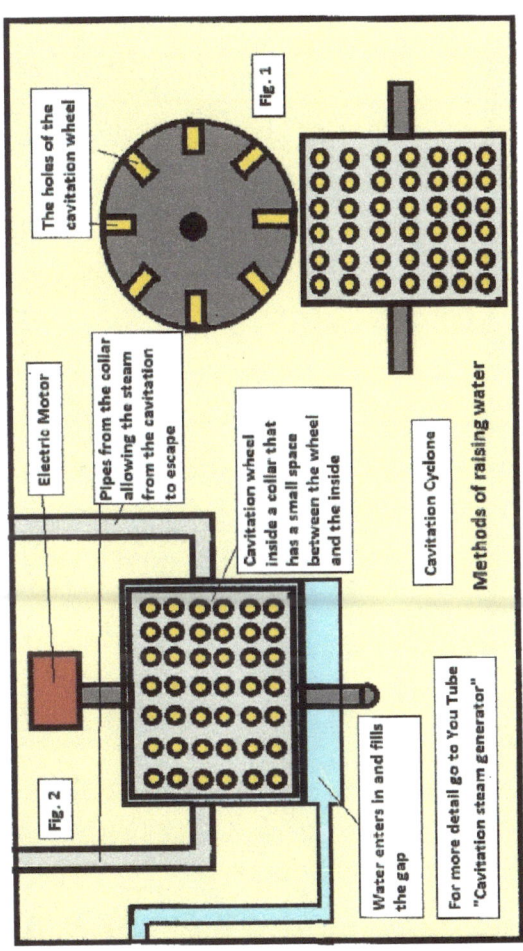

Fig. 1 — The holes of the cavitation wheel

Fig. 2

Electric Motor

Pipes from the collar allowing the steam from the cavitation to escape

Cavitation wheel inside a collar that has a small space between the wheel and the inside

Water enters in and fills the gap

For more detail go to You Tube "Cavitation steam generator"

Cavitation Cyclone

Methods of raising water

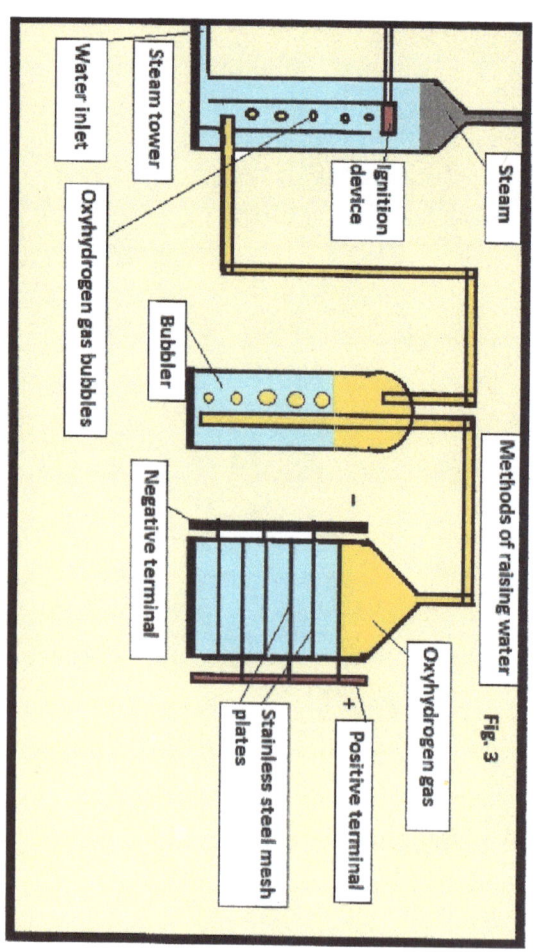

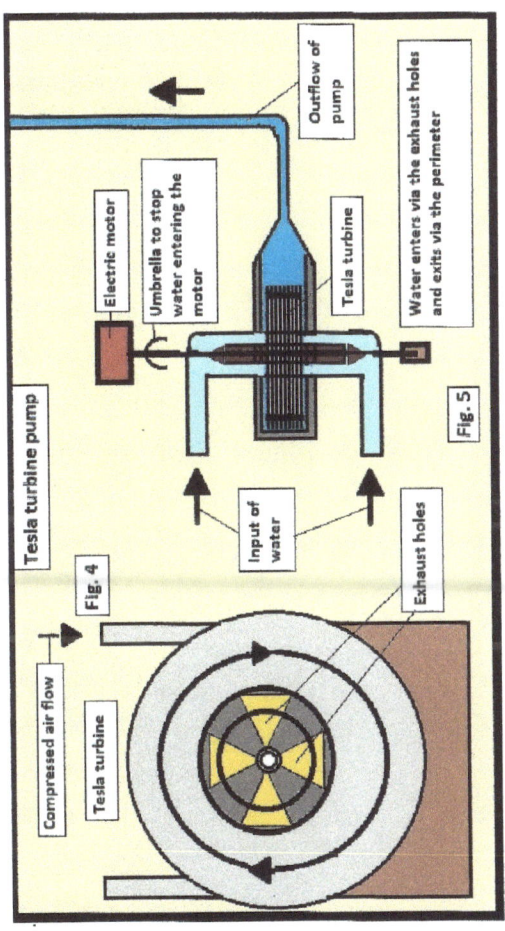

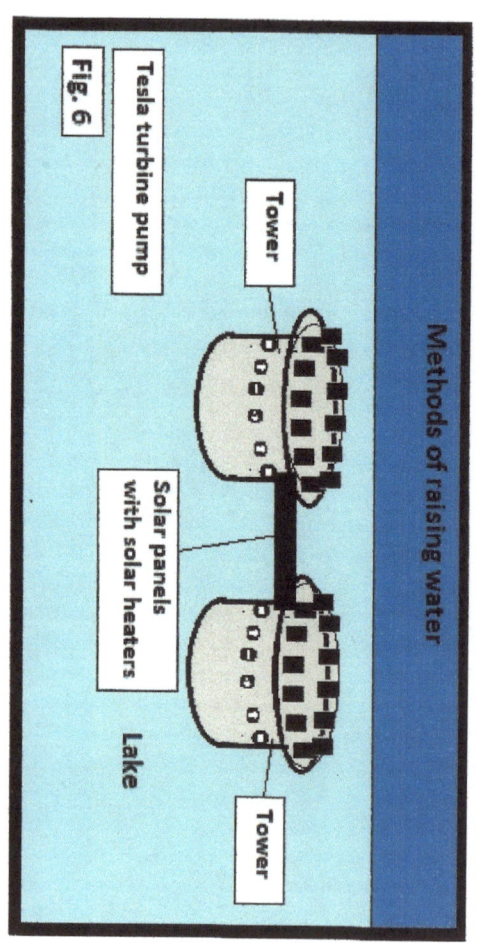

Fig. 6 Tesla turbine pump

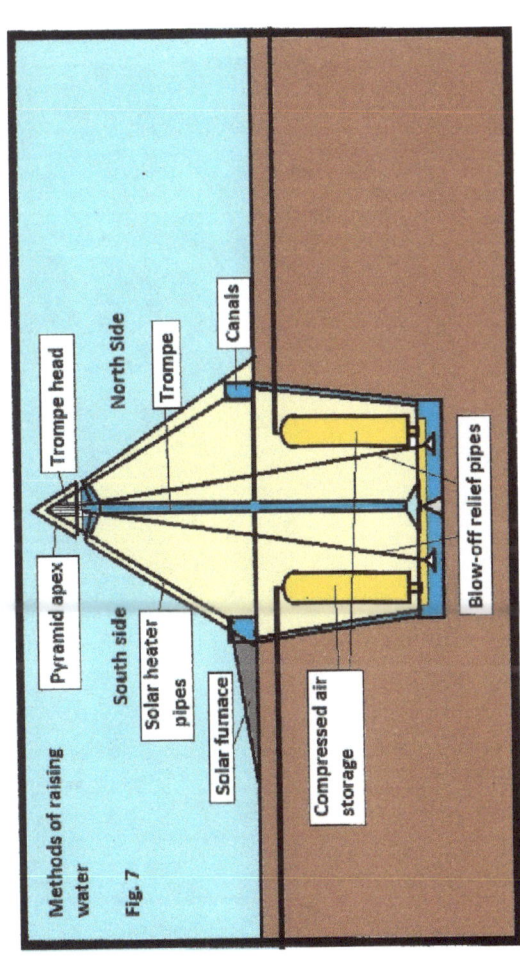

Methods of raising water

Fig. 7

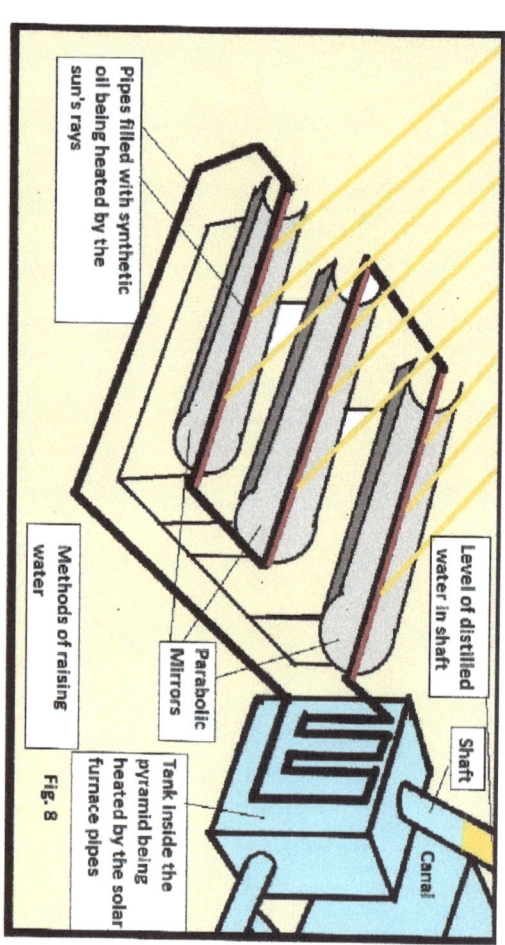

Methods of raising water
Fig. 8

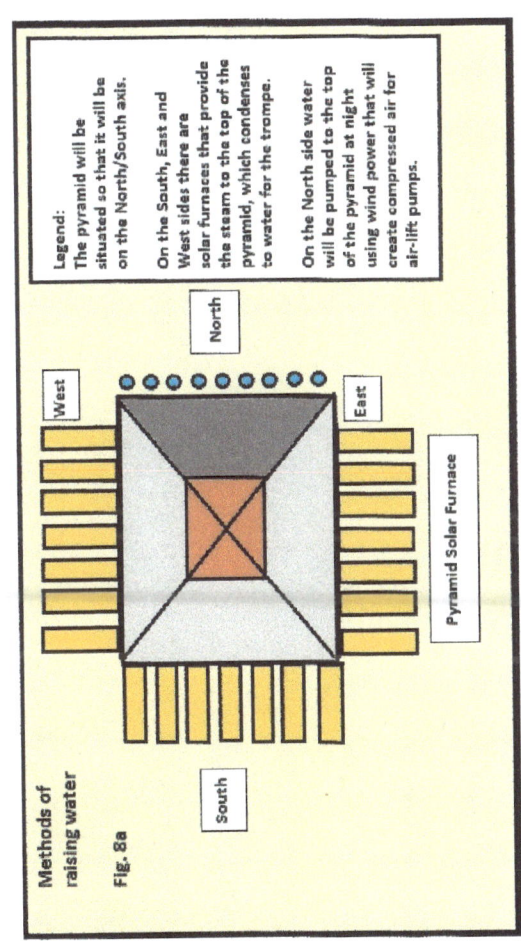

Suffer the Little Children

Methods of raising water – Fig. 9
- Containers
- Containers are connected to the canal via pipes
- Canal
- Steam rises up pipes

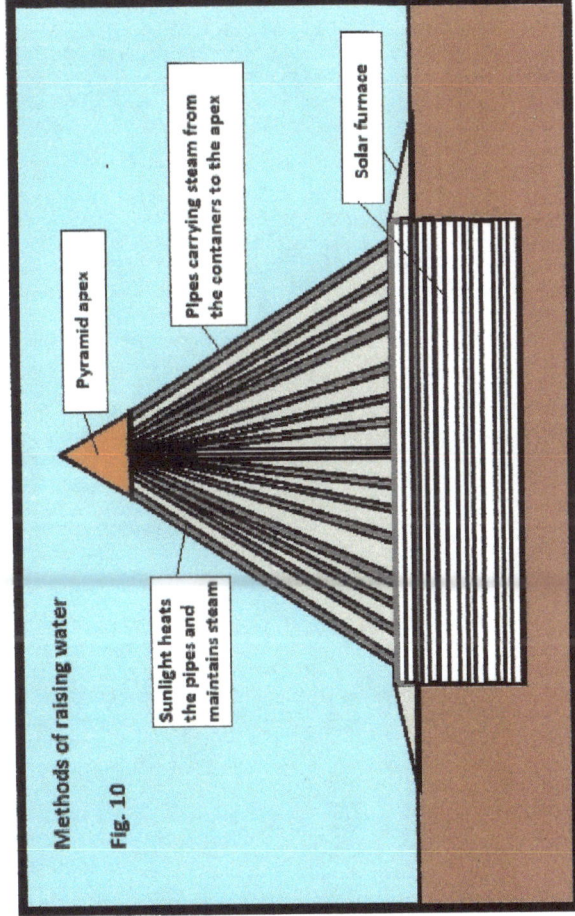

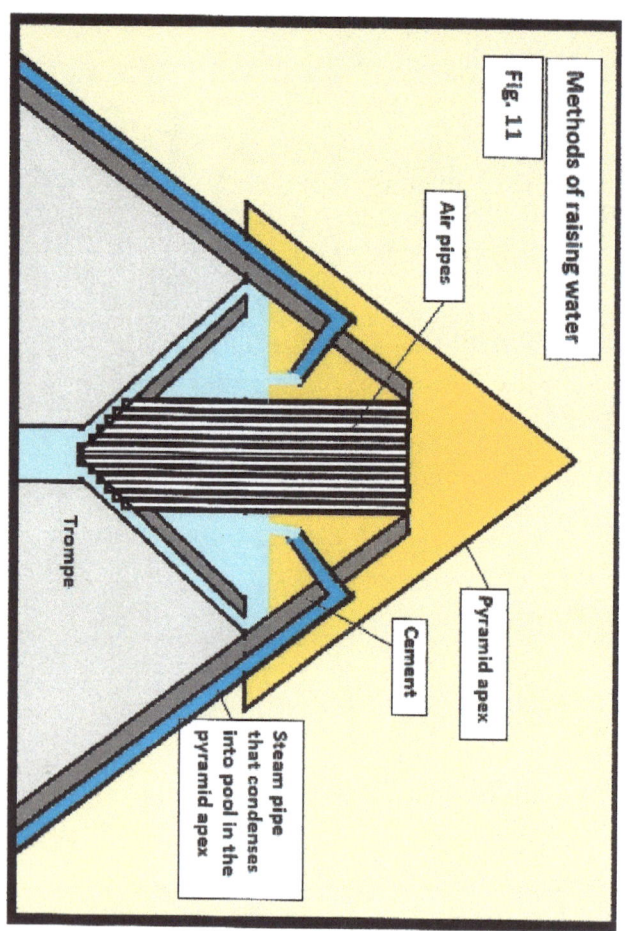

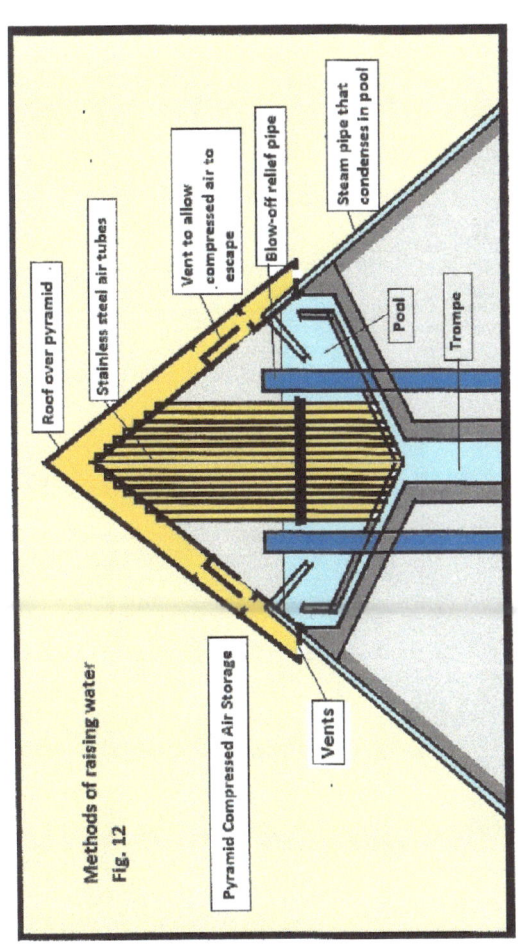

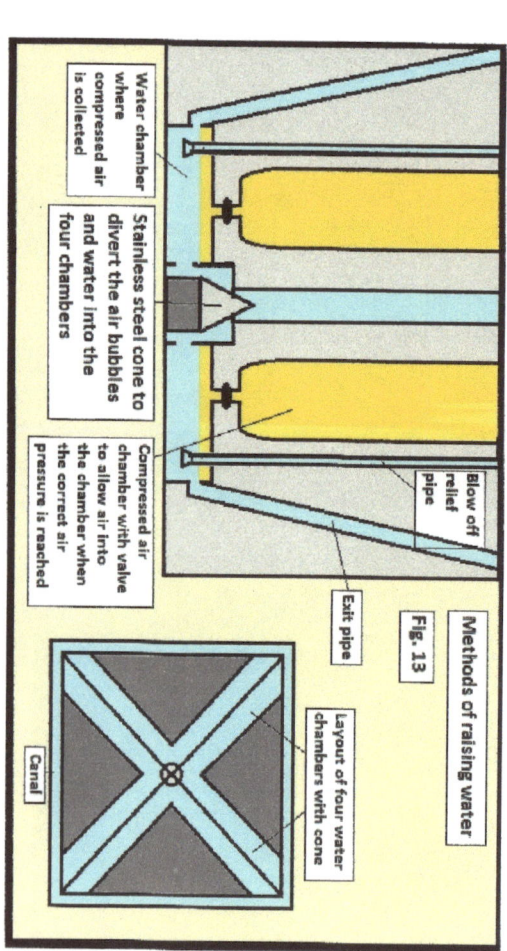

Suffer the Little Children

Summary

When I started this book, my thoughts were turned towards the little children of this world, who come to families throughout the earth that may, through no fault of their own, live in dire conditions. Some may starve after a few days, a few weeks or maybe a year. Some may die for lack of drinking water, or through diseases caused by virus' brought in contaminated water. Some have no clothes in harsh climates, live under governments that are corrupt and deny their people the right to a decent life. Some may even live in areas which still practice slavery.

What ever the circumstances, there are solutions to these children's problems and if mankind would follow the golden rule and love their neighbour as themselves, much of the hardship of the world could be done away. This book does not intend to have all the answers but if the concepts were adopted in our world of today, it would be hoped that at least some children could be spared, allowing them to grow and become useful members of a society that loves them and cares for them.

I hope to have shown that it is possible to have rich soil anywhere in the world; that it is possible to move water, clean water, to anywhere on the earth, even if it needs to flow on level ground. Regardless of the size of the planet's population, adequate housing can be found, even it that means we need to utilize the seas around us. Electrical power should be available to all and at a cost that even the poorest person, wherever they are in this world, should have access to it and should be able to afford it.

The future is in our hands today and if we are to believe our scientists, we need to act soon to save the world from the effects of global climate change. Will our children and grand children come into a polluted, impoverished world or can we make changes today so that future generations can look back at their history and thank us for making wise decisions in their behalf.

Suffer the Little Children

Only time will tell, but if this book helps someone to want to help another child or person then it has accomplished its goal.

Thank you!